AI 绘画时代

Midjourney 用户学习手册

张贤 钟洋 马兆国 编著

人民邮电出版社

北京

图书在版编目（CIP）数据

AI绘画时代：Midjourney用户学习手册 / 张贤，钟
洋，马兆国编著. -- 北京：人民邮电出版社，2023.9
ISBN 978-7-115-62270-9

Ⅰ．①A… Ⅱ．①张… ②钟… ③马… Ⅲ．①图像处
理软件－手册 Ⅳ．①TP391.413-62

中国国家版本馆CIP数据核字(2023)第125021号

内 容 提 要

这是一本系统讲解使用 Midjourney 进行创作的专业教程。全书包括 5 方面的内容，分别讲解了软件的入门基础知识，运用 Midjourney 绘图时的语法、指令，AI 绘画主流的视觉表现及应用类别示例，一些有助于塑造作品风格的艺术家的介绍及其作品赏析，以及进行 AI 绘画创作时应备的一些关键词。本书提供了大量的 AI 绘画案例，可供读者参考。

本书适合插画师、设计师和 AI 绘画创作爱好者阅读与参考。

- ◆ 编　著　张　贤　钟　洋　马兆国
　　责任编辑　王振华
　　责任印制　马振武
- ◆ 人民邮电出版社出版发行　　北京市丰台区成寿寺路 11 号
　　邮编　100164　　电子邮件　315@ptpress.com.cn
　　网址　https://www.ptpress.com.cn
　　北京宝隆世纪印刷有限公司印刷
- ◆ 开本：787×1092　1/16
　　印张：21.5　　　　　　　　2023 年 9 月第 1 版
　　字数：520 千字　　　　　　2023 年 9 月北京第 1 次印刷

定价：129.80 元

读者服务热线：(010)81055410　印装质量热线：(010)81055316
反盗版热线：(010)81055315
广告经营许可证：京东市监广登字 20170147 号

PREFACE
前言

　　人工智能（AI）是当前科技领域的热门话题之一，其发展趋势引起了广泛的关注。随着大数据、深度学习和自然语言处理等技术的不断发展，AI 正逐渐成为第三次工业革命的代表之一。在日常生活中，我们可以看到越来越多的 AI 应用场景，如机器人、人脸识别、无人驾驶等，相关技术正在深刻影响着我们的生活。未来，AI 技术将继续快速发展，拓宽应用场景并创造更多商业价值。

　　在设计工作中，AI 也得到了广泛应用。平面设计、网页设计、UI 设计等领域开始逐渐运用 AI 来提高设计效率。AI 的自动化和智能化特点可以极大地提升设计的质量和效率，帮助设计师在竞争激烈的市场中立于不败之地。

　　本书正是为了帮助广大艺术创作者了解和掌握 AI 绘画技术而编写的。本书将全面介绍 Discord 安装与 Midjourney 注册的方法，以及语法指令的使用技巧。读者可以通过实践将这些技术应用到设计中。此外，本书还贴心地为读者提供了大量的实践案例，让读者在实践中迅速掌握 AI 的应用方法和技巧。

　　本书提供了大量的关键词，可以为读者的 AI 绘画创作做准备；展示的一些艺术家的作品可供读者赏析。笔者相信，通过对本书的学习，读者将会在 AI 的世界中激发新的创作灵感，开拓新的思路，并能够将这些技术应用到实际工作中，使自己的设计达到更高的水平和更好的效果。最后，谢谢读者选择本书，并希望读者在学习完本书后可以获得满意的成果。

提 示

Midjourney 版本更新较快，读者若需要学习了解最新的软件功能和操作方法，可扫描下页的二维码关注"数艺设"微信公众号并输入第 51 页左下角的验证码，根据提示添加班助好友，即可咨询学习过程中遇到的问题或与作者直接交流。

书中的案例作品均配有中英文关键词。为方便读者提高在 Midjourney 中的输入效率，英文关键词（包括缩写、人名、作品名等）均采用正体小写。关键词的中文翻译只表示大概的出图思路和逻辑，仅作参考。

数艺设教程分享

本书由"数艺设"出品，"数艺设"社区平台（www.shuyishe.com）为你提供后续服务。

扫码关注微信公众号

"数艺设"社区平台，为艺术设计从业者提供专业的教育产品。

与我们联系

我们的联系邮箱是 szys@ptpress.com.cn。如果你对本书有任何疑问或建议，请你发邮件给我们，并请在邮件标题中注明本书书名及ISBN，以便我们更高效地做出反馈。

如果你有兴趣出版图书、录制教学课程，或者参与技术审校等工作，可以发邮件给我们。如果学校、培训机构或企业想批量购买本书或"数艺设"出版的其他图书，也可以发邮件联系我们。

关于"数艺设"

人民邮电出版社有限公司旗下品牌"数艺设"，专注于专业艺术设计类图书出版，为艺术设计从业者提供专业的图书、视频电子书、课程等教育产品。出版领域涉及平面、三维、影视、摄影与后期等数字艺术门类，字体设计、品牌设计、色彩设计等设计理论与应用门类，UI设计、电商设计、新媒体设计、游戏设计、交互设计、原型设计等互联网设计门类，环艺设计手绘、插画设计手绘、工业设计手绘等设计手绘门类。更多服务请访问"数艺设"社区平台www.shuyishe.com。我们将提供及时、准确、专业的学习服务。

CONTENTS
目录
———

3

070-287

视觉表现

4

288-305

艺术赏析

5

306-344

关键词灵感

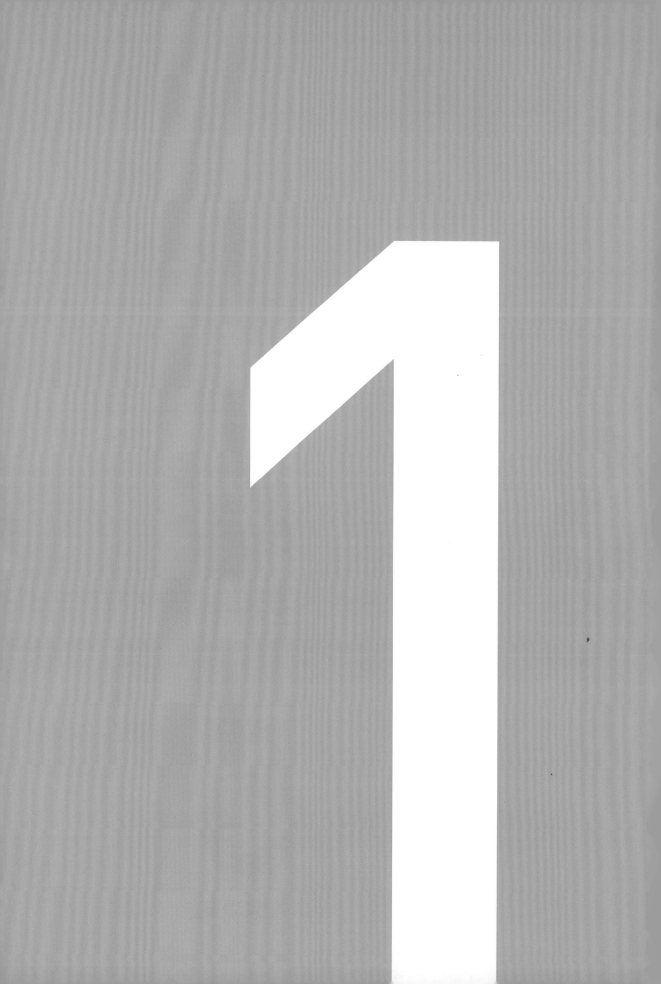

SOFTWARE INTRODUCTION

软件入门

1.1 Discord软件安装

要想使用 Midjourney，首先需要安装 Discord 聊天软件。它们的关系好比微信与小程序，Discord 相当于微信，Midjourney 相当于小程序。大家安装完 Discord 后，在 Discord 中搜索 Midjourney 就可以了。

1.1.1 Discord软件介绍

Discord 是一款流行的多功能聊天软件，它允许用户进行文本、语音和视频聊天。Discord 最初是为游戏社区设计的，现在已经成为各种类型团体的常用沟通工具。Discord 的主要特点如下。

多平台支持：Discord 可以在 Windows、Mac、Linux、iOS 和 Android 等多种平台上使用。

文字聊天：用户可以通过文字消息与其他 Discord 用户交流，这些消息可以在个人或群组聊天中发送。

语音和视频聊天：Discord 具有高质量的语音和视频通信功能，可用于私人或群组通话。

频道管理：Discord 支持创建多个频道，可对不同话题或项目进行分类。

消息历史记录：Discord 可以存储聊天历史记录，允许用户查看以前的消息。

应用程序集成：Discord 可以集成其他应用程序和服务，如 Spotify、Twitch、Midjourney 等。

总之，Discord 是一个非常灵活和易于使用的沟通工具，适用于各种类型的团队和社区。它的功能丰富，用户可以选择使用其中的任何部分，以满足自己的特定需求。

1.1.2 Discord程序下载

打开 Google Chrome 浏览器，输入 Discord 的网址，进入官网首页，如图 1-1 所示。单击"下载"菜单，选择客户端或应用，包括 Mac 版、Windows 版、手机版，或在浏览器右上角打开网页版页面登录或注册，根据操作系统选择相对应的应用程序版本并单击"下载"按钮，如图 1-2 所示。下载完成后，运行安装程序并按照提示完成安装，如图 1-3 所示。本书以网页版在线注册为例进行讲解。

图 1-1

图 1-2

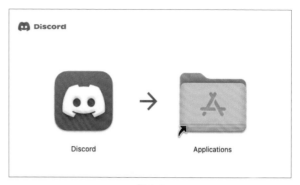

图 1-3

1.1.3 Discord注册与登录

可以通过单击官网导航栏右侧的按钮注册并登录，还可以通过客户端注册并登录，在登录页面左下角"需要新的账号？"右侧单击"注册"，如图 1-4 所示。如果网站页面是英文的，可以使用 Google Chrome 扩展程序对网站进行翻译。

1. 填写信息

填写电子邮箱、用户名、密码、出生日期（尽量选择成年年龄）后，单击"继续"按钮，如图 1-5 所示。勾选"我是人类"选项（这是软件识别你不是机器人的一种手段），如图 1-6 所示；选择相似的图形（如果图片相似性特别强，可以多尝试几遍），单击"检查"按钮，如图 1-7 所示。

图 1-4

图 1-5

图 1-6

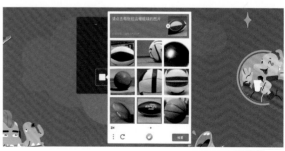

图 1-7

2. 手机验证

单击"使用手机验证"按钮，在弹出的对话框中输入手机号码并单击"发送"按钮，之后根据提示单击相应的图片，最后在弹出的对话框中输入验证码，如图 1-8 ~ 图 1-11 所示。

图 1-8

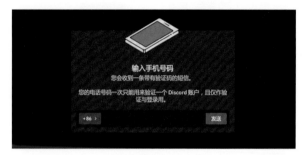

图 1-9

图 1-10

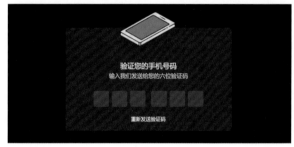

图 1-11

输入密码，单击"确认"按钮后弹出"电话号码已验证！"的提示，完成电话号码验证，如图 1-12 和图 1-13 所示。单击"继续"按钮，进入下一步电子邮箱验证。

图 1-12

图 1-13

3. 电子邮箱验证

在弹出的"需要验证"对话框中单击"使用电子邮件验证"按钮，如图 1-14 所示。如果收不到邮件，可以单击"重新发送验证电子邮件！"按钮，或更改电子邮件地址，之后在弹出的对话框中单击"验证电子邮件地址"按钮，如图 1-15 ~ 图 1-17 所示。完成上述操作后，需要到个人的电子邮箱里验证。

图 1-14

图 1-15

图 1-16

图 1-17

在电子邮箱里完成操作后，Discord 会弹出"电子邮件验证中"对话框，勾选"我是人类"选项，如图 1-18 所示；在弹出的对话框中单击相应的图片继续验证，如图 1-19 所示；在弹出的对话框中再次勾选"我是人类"选项，如图 1-20 所示；最后弹出"电子邮件已验证通过！"的提示，完成注册，如图 1-21 所示。

图 1-18

图 1-19

图 1-20

图 1-21

Discord 软件注册完成后，就可以开始使用了。网页版有自动扩展程序，可以自动翻译界面上的内容，如图 1-22所示。

图 1-22

客户端默认是英文的，如图 1-23所示，需要修改语言设置。单击左下角的设置按钮▣，在语言设置界面将默认语言由英文改为中文，之后单击右上角的关闭按钮完成设置。设置完成后的中文界面如图 1-24所示。

图 1-23

图 1-24

1.2 Midjourney嵌入方法

Midjourney 是一款 AI 绘图工具，只要有关键字，就能通过 AI 算法生成相对应的图片。可以选择不同艺术家的风格进行创作，如凡·高、达·芬奇、塞尚和毕加索等。Midjourney 还能识别特定镜头或摄影术语。将 Discord 安装完成并注册、登录后，还需要将 Midjourney 嵌入 Discord 中，下面讲解具体的嵌入方法。

1.2.1 搜索Midjourney

在 Discord 中单击菜单栏上绿色的浏览器图标，在显示的界面中选择 Midjourney 并将其添加到 Discord 中，如图 1-25 所示。之后弹出欢迎对话框，如图 1-26 所示。单击"Getting Started"按钮进入 Midjourney 的官方频道界面。

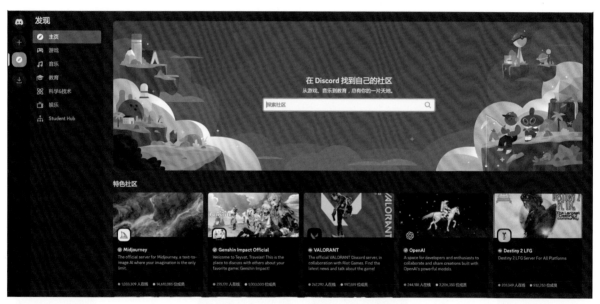

图 1-25

图 1-26

在 Midjourney 的频道
界面中可以看到世界各地
的网友生成的作品，也可
以在频道内发表自己生成
的作品，如图 1-27 所示。

图 1-27

1.2.2 添加服务器

单击菜单栏中绿色的加号图标，弹出"创建服务器"对话框，单击"亲自创建"按钮，如图 1-28 所示。有了个人的服务器，设计出图时就能很直观地观察到效果，下次也能快速找到，避免因群里消息刷新比较快，不容易找到自己的作品，相当于自己建立了一个群。之后单击"仅供我和我的朋友使用"按钮，如图 1-29 所示。

为自己的服务器取个名字，如图 1-30 所示，以便快速找到它。这时就添加了自己的服务器，可以在菜单栏中加号图标的上方看到刚刚创建的服务器，如图 1-31 所示。上述操作完成后，还需要添加一个 Midjourney 机器人才能正常工作。

图 1-28

图 1-29

图 1-30

图 1-31

1.2.3 添加机器人

单击菜单栏中绿色的浏览器图标，在显示的界面上单击 Midjourney，回到 Midjourney 频道界面，如图 1-32 和图 1-33 所示。在界面右上角能够看到好友图标，单击好友图标，界面的右侧会出现好友列表，如图 1-34 所示。再单击 Midjourney 图标，单击"添加至服务器"按钮，如图 1-35 所示。

图 1-32

图 1-33

图 1-34

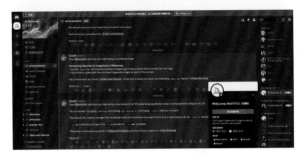

图 1-35

选择之前创建的服务器并单击"继续"按钮，如图 1-36 所示。在弹出的对话框中单击"授权"按钮，如图 1-37 所示。在弹出的对话框中勾选"我是人类"选项，如图 1-38 所示。在弹出的对话框中选择相对应的图片，单击"下一个"按钮后弹出"已授权"提示对话框，如图 1-39 和图 1-40 所示。这样就可以开启 Midjourney 创作之旅了！

图 1-36

图 1-37

图 1-38

图 1-39

图 1-40

1.2.4 订阅注册

　　登录 Midjourney 的官方网站，单击右下角的" Sign In "按钮登录，如图 1-41 所示。在弹出的对话框中单击" 授权 "按钮，如图 1-42 所示。接下来显示的是自己的主页，如图 1-43 所示，这样才能在 Midjourney 官网使用自己的账号。

图 1-41

图 1-42

图 1-43

将网页翻译后，在左侧的菜单栏中单击"管理子系统"按钮，之后显示付费页面。付费形式分为两种：一种为按年计费，另一种为每月结算，如图 1-44 和图 1-45 所示。可以根据自身情况选择相应的付费形式。每月结算的形式存在自动续费功能，如果下个月不想继续使用，需要单独取消。支付页面支持信用卡支付，如图 1-46 所示。

图 1-44

图 1-45

图 1-46

SYNTAX
COMMANDS

语法指令

2.1 Midjourney基础入门

本节主要讲解如何查看用户信息、设置和预设，以及各种指令的使用方法，出图时如何使用恰当的后缀参数才能生成更加精准的图片。

2.1.1 指令参数合集

指令和参数是 Midjourney 的核心，理解并掌握不同的指令和参数，才能真正掌握 Midjourney。下面将所有常用指令和参数大致分为 3 类，即前缀指令、后缀参数和 Emoji 功能。一起来了解一下 Midjourney 的指令和参数吧。

1. 前缀指令

用户可以通过前缀指令与 Discord 中的 Midjourney Bot 进行交互。前缀指令通常用于创建图像、更改默认设置、监视用户信息，以及执行其他有用的任务。用户可以选择使用其中的任何部分，以满足特定的创作需求。

前缀指令	
/ask	得到一个问题的答案
/blend	将 2~5 幅图融合在一起，生成新的图像，不支持文本提示
/describe	以图生文。将自己的图像上传，系统会根据图像生成 4 项描述，供参考使用
/fast	切换到快速模式，出图速度快（会员默认使用快速模式）
/relax	切换到放松模式，出图速度慢
/help	显示有关 Midjourney Bot 的有用基本信息和提示
/imagine	最基础的绘画指令。在 imagine 后面输入关键词即可进行 AI 绘画
/info	查看有关账户及任何排队或正在运行的信息
/stealth	隐身模式。防止生成的图像在 Midjourney 网站上被其他人看到
/public	公开模式。默认为开放社区，所有生成的图像都可以在 Midjourney 官网上被其他人看到
/subscribe	生成订阅，直接跳转到订阅服务
/settings	查看和调整 Midjourney Bot 的设置
/prefer option	创建或管理自定义选项
/prefer option list	查看当前的自定义选项
/prefer suffix	指定要添加到每个提示词末尾的后缀
/show	使用图像的 Job ID，然后在 Discord 中重新生成图像
/remix	Remix 模式是一项实验性功能，可能会随时更改或删除。作用是在生成好的原有图像的基础上进行各种风格的调整
::	生活中有很多词的意思 AI 是无法正常理解的，这时就需要一个分割符号将词分开，便于 AI 理解。例如，想生成一条很热的狗，但是输入 hot dog 后会生成一个热狗。可以在词的中间加上 "::"，使之变成 hot:: dog，这样就会生成一条很热的狗了
{}	在大括号 {} 内分隔选项列表，可以快速创建和处理多个提示词变体，包括文本提示词、图像提示词、参数或提示词权重

弃用指令
/private（替换为"/stealth"）
/pixels
/idea

2. 后缀参数

后缀参数是添加到提示中的选项，可更改图像的生成方式。使用后缀参数可以更改图像的长宽比例，可以切换 Midjourney 的模型版本，也可以更改使用的 Upscaler 等。

> 注：许多苹果设备会自动将双连字符（--）更改为一字线（—），Midjourney 都可以识别。

基本参数	
--aspect 或 --ar	设置画幅比例，格式为 --ar x:y，如 --ar 3:4 或 --ar 9:16
--chaos 或 --c	--chaos 接受值为 0~100。--chaos 默认值为 0 参数值越小，图像效果越符合描述指令；参数值越大，图像越具有意想不到的构图或艺术效果
--no	否定关键字，即不希望图像中出现这个关键字对应的事物。例如，不希望画面中出现植物，那就输入 --no plants，这样图像中就不会出现植物元素了
--quality 或 --q	--q 0.5，--q 1，--q 2，数值越高，画面质量越好，所消耗的时间就越长
--seed	如果已经生成一幅满意的图像，但是后期又想针对这幅图像做细节上的优化和调整，就需要用到"--seed"；获取相应的 Seed 值后，微调描述词即可。系统默认的 Seed 值是 0~4 294 967 295 范围内的一个随机数值
--sameseed	使用相同的种子来生成图片，将会生成非常相似的图片（仅应用于 V1~V3 版本）
--stop	--stop 10~100，在图像由 0 到 100% 生成的过程中，选择一个节点停止生成，以较早的百分比停止会产生更模糊、更不详细的效果。如果想要模糊且抽象的图像效果，不妨试一下
--style	4a、4b、4c 可以在 Midjourney V4 版本中切换
--stylize 或 --s	可以影响默认的 Midjourney 美学风格，数值为 0~1000，默认值为 100；数值不同，美学风格也不同

模型版本参数	
--hd	增加清晰度，适合用作抽象图片和风景图片的后缀
--test	通用艺术模型
--testp	照片写实模型
--version 或 --v	Midjourney 的不同算法。目前有 5 个版本，即 V1、V2、V3、V4、V5，当前算法 (V4) 是默认设置
--niji	专注于二次元动漫风格的模式，最新为 V5 版本

> 注：Midjourney 会定期发布新模型版本，以提高效率和质量。不同的模型擅长处理不同类型的图像。

其他参数	
--creative	Midjourney 偶尔会临时发布新模型，以供社区测试和反馈。目前有两种可用的测试模型：--test 和 --testp，它们可以与 --creative 参数结合使用，以获得更多不同的成分
--iw	用于设置文本提示与上传的图片之间的权重比例
--video	保存演算过程视频。此功能仅限于 V1、V2、V3 版本
--tile	该后缀可以生成四方连续的重复拼贴图像，适合用来制作织物、壁纸和纹理的无缝图案

弃用参数	
--width 和 --w	替换为 --aspect
--height 和 --h	替换为 --aspect
--fast	替换为 --quality
--vibe	现在称 V1
--hq	
--newclip	
--nostretch	
--old	
--beta	

3.Emoji 功能

Emoji 功能	
✉️	与信封表情符号互动，给完成的图像发送✉️，会收到 Midjourney 发来的消息，内容包括图像的种子编号和 Job ID
★	星星符号会将图片标记为"收藏"。可将图像发送到公用 #favorites 频道，并在官网上添加到自己的书签中
✖	随时取消或删除工作。使用✖表情符号对图像做出反应时，Midjourney 网站会删除该图像。如果想从 Midjourney 网页中删除一幅图片，但在 Discord 中找不到该图片，可使用 /show 指令恢复图像

2.1.2 用户信息

/info 指令可用于查看有关当前排队和正在运行的作业、订阅类型、续订日期等信息。

使用 /info 指令：在频道下方输入 /info 并按 Enter 键，会弹出信息窗口，如图 2-1 和图 2-2 所示。

图 2-1

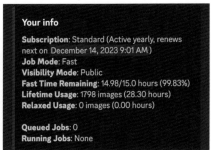

图 2-2

· Subscription（订阅）

订阅部分显示用户订阅了哪个套餐，以及下一个续订日期。

· Job Mode（作业模式）

显示当前是处于快速模式还是放松模式，放松模式仅适用于 Standard 和 Pro Plan 订阅者。

· Visibility Mode（可见模式）

显示当前是处于公开模式还是隐身模式，隐身模式仅适用于 Pro Plan 订阅者。

· Fast Time Remaining（快速剩余时间）

显示当月剩余的快速 GPU 时间，快速 GPU 时间每月重置一次，不会结转。

· Lifetime Usage（终身使用）

显示用户一生的中途统计数据，生成的图像包括所有类型。

· Relaxed Usage（轻松使用）

显示用户当月放松模式的使用情况，重度放松模式用户的排队时间会稍微长一些，放松模式的使用量每月重置。

· Queued Jobs（排队作业）

列出所有排队等待运行的作业，最多可以同时允许 7 个作业排队。

· Running Jobs（运行作业）

列出当前正在运行的所有作业，最多可以同时运行 3 个作业。

2.1.3 设置和预设

/settings 指令通常为模式版本、样式值、质量值和升级器版本等常用选项提供切换按钮。设置也有公开模式和隐身模式命令的切换按钮。

使用 /settings 指令：在频道下方输入 /settings 并按 Enter 键，会弹出设置面板，如图 2-3 和图 2-4 所示。

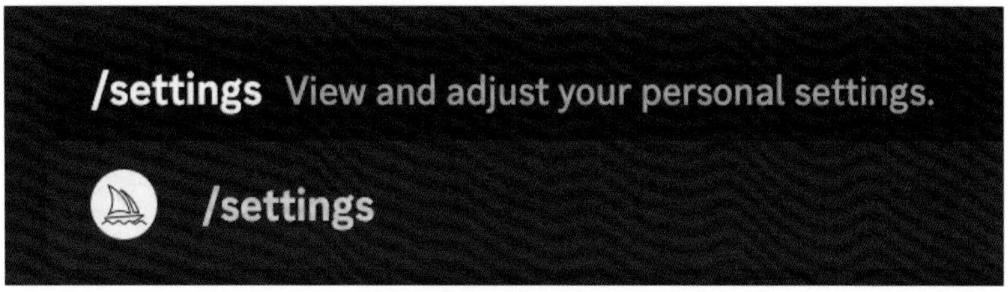

/settings View and adjust your personal settings.

/settings

图 2-3

图 2-4

·**MJ version 1、MJ version 2、MJ version 3、MJ version 4、MJ version 5 、MJ version 5.1、RAW Mode**

这几个版本均为正常出图版本。

·**Niji version 4、Niji Version 5**

Niji version 4 和 Niji version 5 为二次元漫画风格的出图版本。

·**Stylize low、Stylize med、Stylize high、Stylize very high**

用于设置作品的风格化参数。风格低 = --s 50，风格中 = --s 100，风格高 = --s 250，风格非常高 = --s 750。

·**Public mode**

在公开模式和隐身模式之间切换，对应 /public 和 /stealth 指令。开启为公开模式，关闭为隐身模式。

·**Fast mode**

在快速模式和轻松模式之间切换，对应 /fast 和 /relax 指令。开启为快速出图模式，关闭为慢速出图模式。

·**Remix mode**

切换到混音模式，这是一项实验性功能，可能会随时更改或删除。

·**Reset Settings**

返回默认设置。

快速模式是有固定时长的，购买会员本质上就是购买快速出图的时长；不同的会员快速出图的总时长不同。例如，月会员出图总时长为 15 小时。当快速出图的时长用完后，就会变成放松模式出图，速度相对较慢，但是出图的次数是无限的。快速出图的时长用完了，也可以再额外购买更多的快速出图时长，如图 2-5 所示。

图 2-5

注：出一幅图的时长由图片的尺寸、质量等因素决定。Midjourney Bot 机器人平均需要大约一分钟的 GPU 时间来创建图像。

· 自定义首选项

使用 /prefer 指令创建自定义选项，可以自动将常用参数添加到提示词末尾。

/prefer auto_dm：完成的工作会自动发送到直接消息。

/prefer option：创建或管理自定义选项。

/prefer option list：查看当前的自定义选项。

/prefer suffix：指定要添加到每个提示词末尾的后缀。

· 偏好选项

/prefer option set ⟨name⟩ ⟨value⟩：创建可将多个参数快速添加到提示词末尾的自定义参数，如图 2-6 和图 2-7 所示。

/prefer option set mine --hd --ar 7:4：创建一个名为"我的"的选项，转换为 --hd --ar 7:4。

将值字段留空，以删除选项。

/prefer option list：列出创建的所有选项，如图 2-8 所示。用户最多可以有 20 个自定义选项。

要删除自定义选项，可使用 /prefer option set ⟨name to delete⟩ 值字段并将其留空。

图 2-6

图 2-7

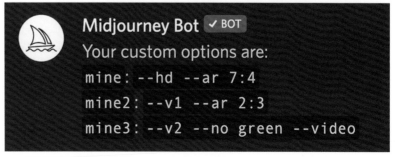

图 2-8

2.1.4 出图指令

为了更好地完成最终的图片效果，本小节将通过基础出图、垫图技巧、图像混合、以图生文、多种提示、排列提示等方法来帮助大家完成出图。

1. 基础出图

/imagine 指令是用于创建图像、更改默认设置、监视用户信息，以及执行其他有用任务的指令，使用 /imagine 指令基础出图的步骤如下。

（1）在指令框中键入 /imagine 或选择 /imagine 指令。

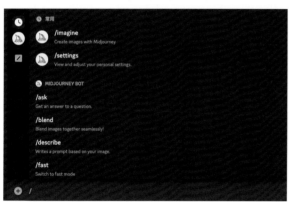

图 2-9

> 注：如果输入指令时没有看到弹出的 /imagine 指令，请多尝试几次，或者将 Midjourney 机器人重新添加到你的服务器中，如图 2-10 所示。

图 2-10

（2）在 prompt 后方输入要创建的图像的描述词。

图 2-11

> 注：描述词中禁止出现血腥、色情等敏感性词汇。

（3）按 Enter 键发送消息。

图 2-12

> 注：初次发送指令时，Midjourney Bot 将生成一个弹出窗口，要求用户接受服务条款。在生成任何图像之前，用户必须先同意服务条款。

（4）等待大约一分钟的时间，Midjourney 就可以生成 4 幅风格迥异的图像了。

图 2-13

（5）单击图像下方的按钮，会生成不同的图像效果。单击 按钮，系统将按照原始指令重新生成 4 幅新的图像。

图 2-14

单击图像所对应的 U 按钮，可以在该图像基础上生成更高清、细节更丰富的图像。

图 2-15

单击图像所对应的 V 按钮，系统会在该图像基础上重新生成整体风格与之类似但细节略有不同的 4 幅新图像。

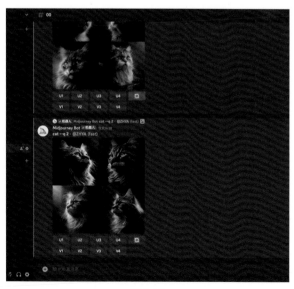

图 2-16

（6）评价图像。单击 U 按钮，得到放大的图像后，将出现一组新选项。

Make Variations：创建放大图像的变体并生成包含 4 个选项的图像。

Beta/Light Upscale Redo：使用不同的升级器模型重新升级。

Web：跳转到 Midjourney 网页上打开图库中的图像。

Favorite：给该作品点赞。

图 2-17

在 Midjourney 网页上打开图库中的图像后，单击图 2-18 中的■ 按钮，可以调出多个选项。

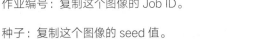

注：这里为了确保大家可以看清楚，将网页翻译成了中文。

· **复制**

全命令：复制这个图像的所有命令，包括图像描述与后缀参数。

迅速的：只复制图像描述，不复制后缀参数。

作业编号：复制这个图像的 Job ID。

种子：复制这个图像的 seed 值。

图 2-18

· **打开**

新标签：在浏览器的新页面中打开。

在 Discord 中打开。

图 2-19

图 2-20

· 用于

    头像：可将此图像作为自己账号的头像。

    简介封面：可将此图像作为账号的封面图。

图 2-21

· 报告

    报告：给差评的意思。

    不要给我看这个：不喜欢的意思。

· 添加到收藏

    添加到收藏：将此图像添加到收藏。

    最喜欢的：为此图像点个赞。

· 保存图片

    将此图像保存到本地。

图 2-22

可以在 Midjourney 网站或 Discord 上单击一个笑脸按钮来评价任何放大的图像，如图 2-23 所示。每天排名前 1000 名的图像评分者都会获得一小时的免费快速模式时间。

图 2-23

2. 垫图技巧

使用 /imagine 指令出图的方法简单易操作，但有时出的图与我们脑海中想要的大相径庭，这时就要掌握另一种出图方法——"垫图"，也称"喂图"。所谓垫图，就是将喜欢的图像放到 /imagine 提示词中，作为提示词的一部分，来影响作品的构图、风格和颜色。图像提示词可以单独使用，也可以与文本提示词一起使用，尝试将不同风格的图像组合起来，可以获得令人意想不到的效果。

想将图像添加到提示词中，可以将自己的本地图片上传到 Discord 中：复制链接，再将其粘贴到 /imagine 提示词中。图片链接必须以 .png、.gif 或 .jpg 等扩展名结尾。添加图像地址后，再添加其他的文本和参数，以完成提示词的添加。

· 具体操作步骤

（1）单击"上传文件"按钮，在打开的文件夹中选择要上传的图片，然后单击 Open（打开）按钮上传图片。

图 2-24

图 2-25

（2）将图片上传至 Discord 后，按 Enter 键。

图 2-26

图 2-27

（3）单击鼠标右键，在弹出的快捷菜单中选择"复制链接"选项。

图 2-28

（4）将复制的链接粘贴到 /imagine 提示词中。添加图像地址后，一定要按 Space 键，再添加其他文本和参数，这里输入 vector illustration of a dog。

图 2-29

（5）按 Enter 键，得到图 2-30 所示的效果。通过对比图 2-30 和图 2-31 中狗的形象可以发现，生成的图像与之前的原图相似度是比较高的。

图 2-30

图 2-31

如果不垫图，而是直接输入 vector illustration of a dog，那么得到的将会是图 2-32 所示的效果。由此可见"垫图"的作用与优势。

图 2-32

> 注：图片 URL 必须位于文本提示词的前面；提示词必须包括一幅或两幅图片的 URL，以及文本提示词；图片 URL 与后面的文本提示词之间必须加空格。

·图像权重参数

有时候即便垫了图，生成的图像还是与原图有较大的区别，这时候就要用到一个后缀参数"--iw"。"--iw"通常在垫图时使用，用于设置文本提示词与上传的图片之间的权重比例，较高的 --iw 值意味着图片对生成的作品产生的影响更大；相反，--iw 值越小，图片对生成的作品产生的影响就越小。

图像权重参数（从左到右，由低到高）				
--iw .5	--iw .75	--iw 1（默认）	--iw 1.5	--iw 2
图像权重低 文本提示词权重高		图像权重与 文本提示词权重持平	图像权重高 文本提示词权重低	

还是拿小狗的图片举例，当在 /imagine 提示词中添加了图 2-33 所示的链接和文本描述词后，只需要在后面加上"--iw"参数即可，生成的图像如图 2-34 所示。

图 2-33

图 2-34

完整的垫图语法结构为：/imagine prompt: 图片链接 （空格） 文本描述 （空格） --iw 数值。--iw .5 生成的图像如图 2-35 所示，--iw 2 生成的图像如图 2-36 所示。

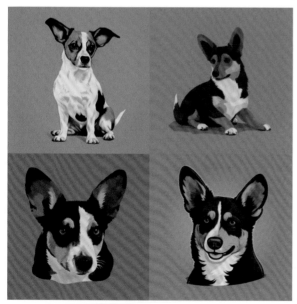

图 2-35

图 2-36

3. 图像混合

/blend 指令可以将 2~5 幅图像通过算法融合为一幅新的图像，具体步骤如下。

（1）输入 /blend 并按 Enter 键，系统会提醒添加图片（2~5 幅）。

> 注：/blend 最多可处理 5 幅图像；/blend 不支持文本提示词；为获得最佳效果，请上传与想要的图像具有相同宽高比的图像。

图 2-37

（2）选择需要添加的图片，并单击 Open（打开）按钮。

图 2-38

（3）将图片上传完毕后，按 Enter 键。

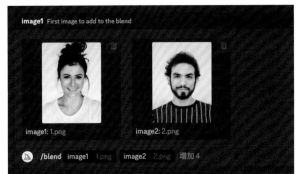

图 2-39

（4）系统开始处理图像。

图 2-40

（5）生成图像。

图 2-41

4. 以图生文

很多新手在出图的时候脑海中总是一片空白，不知道该输入哪些描述词和指令参数；或者看到一幅特别精美的作品，自己也想模仿一幅，却不知道图中的描述与关键词有哪些。如果能学会使用 /describe 指令，这些问题将迎刃而解。/describe 指令是用于识别图片中的文本描述及关键词的指令。只需要将自己喜欢的图片通过 /describe 指令上传到 Discord 中，系统便会生成 4 组文本描述供上传者参考。只需要对相应关键词稍作修改，就会得到一组可以媲美原图效果的作品。具体步骤如下。

（1）准备好一幅参考图。注意，禁止使用血腥、暴力、色情等违规的图片。

图 2-42

（2）输入 /describe 指令，按 Enter 键。

图 2-43

（3）系统提醒上传图片，选择需要上传的图片，并单击 Open（打开）按钮。

图 2-44

（4）将图片上传完毕后，按 Enter 键。

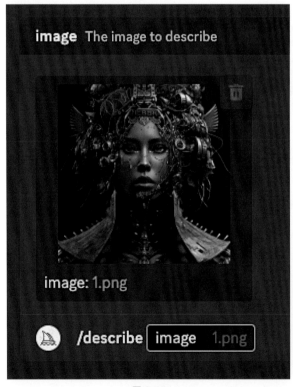

图 2-45

（5）稍等片刻，便会得到 4 组关键词，单击图片下方的数字按钮（1、2、3、4 中的任意一个），系统会根据生成的对应关键词生成图片。

图 2-46

如果出现弹窗提醒，请直接单击"提交"按钮，也可以修改关键词后再单击"提交"按钮。

图 2-47

（6）系统生成 4 幅风格与原图接近的图片。

图 2-48

当然，也可以将生成的关键词复制出来，重新进行修改和调整，并再次通过 /imagine 指令生成新的图像，如图 2-49 所示。只要关键词准确，生成的图片效果就会无限接近原图效果，甚至超过原图效果。

图 2-49

修改和调整后的关键词如下。

关键词组合

hyper realistic mixed media portrait of a beautiful mechanical steampunk gundam robot, medusa, grandiose portraits , symmetrical head, metal gear, aztec art, post process, 4k, highly ornate intricate details, in the style of horizon zero dawn

一个美丽的机械蒸汽朋克高达机器人的超逼真混合媒介肖像，美杜莎，宏伟的肖像，对称的头部，金属齿轮，阿兹特克艺术，后期处理，4K，高度华丽复杂的细节，《地平线：零之曙光》（游戏名）的风格

5. 多种提示

生活中有很多词的意思 AI 是无法正常理解的，这时就需要一个分割符号将词分开，便于 AI 理解。在提示词中添加双冒号"::"，Midjourney Bot 就会分别考虑每个提示词的意思，而非整体的意思。例如，如果想生成一栋白色的房子，那就需要在白色和房子之间加上"::"，变成 white:: house，这样就会生成一栋白色的房子而不是白宫。再以热狗（hot dog）为例，如果想生成一条很热的狗，那就可以改成 hot:: dog，以此类推。

注：双冒号"::"之间没有空格。

直接输入 The white house，系统会理解为白宫。输入 The white:: house，系统则会直接理解为白色的房子，生成的效果如图 2-50 所示。

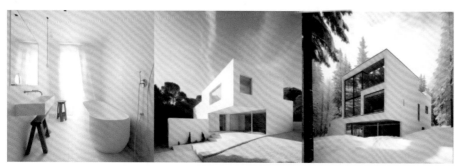

图 2-50

直接输入 hot dog illustration，系统会理解为一幅热狗插画并生成 4 幅图，如图 2-51 所示。如果输入 hot:: dog illustration，系统则会生成图 2-52 所示的一组温暖的、与狗相关的插画。

图 2-51　　　　　　　　　　　　　　　　　　图 2-52

如果输入 hot dog:: illustration，系统会生成一组热狗插画，如图 2-53 所示。

图 2-53

· 提示词权重

当使用双冒号 "::" 将提示词分成不同的部分时，也可以在双冒号后添加一个数字，以强调这部分的相对重要性。例如，输入 hot:: dog illustration，可以生成一条比较温暖的狗。如果更改为 hot::3 dog illustration，代表 3 倍的热量，可以生成一条非常热的狗，如图 2-54 所示。

图 2-54

· 后缀参数" --no "

这里需要介绍一个后缀参数" --no ",它是一个否定关键字,即不希望图像中出现这个关键字对应的事物。例如,不希望画面中出现植物,那就输入 --no plants,这样图像中就不会出现植物元素了。

6. 排列提示

排列提示功能用于创建多个任务,可以提高出图效率,但仅限于快速出图模式下使用。可以通过排列提示创建 /imagine 提示词中的任何组合和排列,包括文本提示词、图像提示词、参数或提示词权重。

· 排列提示基础

在"{ }"内分隔列表,可以快速创建和处理多幅图像。

例如,在 /imagine prompt 的后面输入 a {red, green, yellow} flower(一朵 { 红色,绿色,黄色 } 的花),其实就等于输入 3 个指令,分别是 /imagine prompt a red flower(一朵红色的花),/imagine prompt a green flower(一朵绿色的花),/imagine prompt a yellow flower(一朵黄色的花),那么系统可以同时生成 3 组图像,如图 2-55~ 图 2-57 所示。说得直白一点,就是输入一个指令可以生成 3 组图像。

图 2-55　　　　　　　　　　图 2-56　　　　　　　　　　图 2-57

除了可以排列不同的内容,还可以排列不同的后缀参数,如不同的比例 --ar 和不同的版本 --v。

例如,在 /imagine prompt 的后面输入 a very beautiful rose --ar {3:2, 1:1, 1:2}(一朵非常漂亮的玫瑰花 { 3：2 比例,1：1 比例,1：2 比例 }),其实就等于输入 3 个指令,分别是 /imagine prompt a very beautiful rose --ar 3:2,/imagine prompt a very beautiful rose --ar 1:1,/imagine prompt a very beautiful rose --ar 1:2,生成的效果如图 2-58~ 图 2-60 所示。

图 2-58　　　　　　　　　　　图 2-59　　　　　　　　　　　图 2-60

又如，在 /imagine prompt 的后面输入 a very cute girl -- {v 4, niji 5}（一个非常可爱的女孩 { v4 版本，niji 5 版本 }），其实就等于输入两个指令，分别是 /imagine prompt a very cute girl --v 4，/imagine prompt a very cute girl --niji 5，生成的效果如图 2-61 和图 2-62 所示。

图 2-61　　　　　　　　　　　　　　　　　图 2-62

除此之外，还可以使用多重和嵌套排列形式。在 /imagine prompt 的后面输入 a {red, green} bird in the {jungle, desert}（一只 { 红色，绿色 } 的鸟在 { 丛林，沙漠中 }），其实就等于输入 4 个指令，分别是 /imagine prompt a red bird in the jungle（一只红色的鸟在丛林中），/imagine prompt a red bird in the desert（一只红色的鸟在沙漠中），/imagine prompt a green bird in the jungle（一只绿色的鸟在丛林中），/imagine prompt a green bird in the desert（一只绿色的鸟在沙漠中），生成的效果如图 2-63~ 图 2-66 所示。

图 2-63　　　　　　　　　　　　　　　　图 2-64

图 2-65　　　　　　　　　　　　　　　　图 2-66

　　类似这样的排列形式还有很多种，只需要发挥想象，就可以大幅提升工作效率。

2.1.5 /show指令

　　可以使用 /show 指令来恢复丢失的作品，或刷新旧作品，以产生新的变化，升级或使用更新的参数及功能。

注：/show 指令只适用于自己的作品。

Job ID 是一串随机生成的数字代码，如 9333dcd0-681e-4840-a29c-801e502ae424，它是 Midjourney 生成的每个图像的唯一标识符，就像每幅图像的身份证一样。

Job ID 通常可以在网站的个人主页上、图像对应的网页链接中或者图像的文件名中找到。

1. 查找 Job ID

查找 Job ID 有以下 3 种方式。

一是可以在网站的个人主页上单击■按钮，然后执行"复制 > 作业编号"命令来复制 Job ID，如图 2-67 所示。

图 2-67

二是在图像对应的网页链接中找到该图像的 Job ID，如图 2-68 所示。

图 2-68

三是在图像的文件名中查看 Job ID。例如，一幅图像的文件名为 HYA_cat_2b8b598b-1123-4f21-87ce-e4a9ff4a7848，那么2b8b598b-1123-4f21-87ce-e4a9ff4a7848 就是它的 Job ID，如图 2-69 所示。

图 2-69

2. 实际应用

如果在 Discord 中不小心将之前生成的图像消息删除了，可以在网站的个人主页上单击■按钮，然后执行"复制 > 作业编号"命令来复制 Job ID，然后在任意频道中输入 /show job_id，如图 2-70 所示。

图 2-70

将之前复制的 Job ID 粘贴到 job_id 后面，并按 Enter 键，这样就可以恢复这幅图像了，如图 2-71 和图 2-72 所示。

图 2-71

图 2-72

3. 扩展延伸

单击图像右上方的表情图标，找到信封表情，并单击发送，如图 2-73 所示。

图 2-73

在左边的频道中，Midjourney Bot 会发来一条消息，如图 2-74 所示。打开消息，会看到该图像更详细的内容：Job ID 和 Seed 值，如图 2-75 所示。

图 2-74

图 2-75

2.1.6 /remix指令

Remix 模式可以允许在原作品的基础上添加额外的提示词，改变原有画作的一切内容，如光源、主体、风格等。

> 注：Remix 模式是一项实验性功能，可能会随时更改或删除。

使用 /remix 指令的步骤如下。

（1）使用 /settings 指令调出设置，然后单击 Remix mode 按钮。

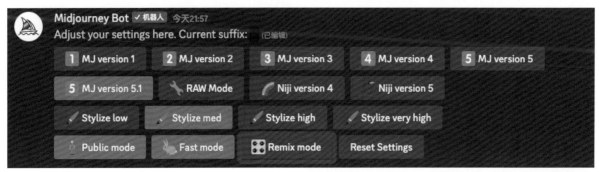

图 2-76

（2）在 /imagine 指令框中输入 basketball --v 5，单击"提交"按钮。

图 2-77

（3）单击图像下方的 V1、V2、V3、V4 中的任意一个按钮，会弹出一个窗口，可以在其中修改或添加关键词，单击"提交"按钮，系统会生成新的图像。

图 2-78

如果单击 V1 按钮，并将关键词 basketball 改为 basketball fire，可以得到图 2-79 所示的一组图像。

用同样的方法还可以将关键词 Cute boy with short hair, full body 改为 Cute boy with black hair, cartoon, 修改前后的图像如图 2-80 所示。

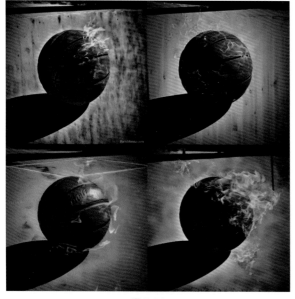

图 2-79

图 2-80

总之，Remix 模式允许在原有画作的基础上做各种形式的修改和调整。读者可以自行尝试，也许会得到更多意想不到的效果。

2.1.7 后缀参数

后缀参数是添加到提示词中的选项，可更改图像的生成方式，用"-- ＋ 后缀"来表示。后缀参数可以更改图像的宽高比、切换模型版本等。后缀参数总是添加到文本提示词的末尾，可以为文本提示词添加多个后缀参数，如图 2-81 所示。

 /imagine prompt　a vibrant california poppy --aspect 2:3 --stop 95 --no sky

图 2-81

注：许多苹果设备会自动将双连字符（--）更改为破折号（—），Midjourney 两者都接受。

1.--ar

　　--ar 是一个用来设置画幅比例的后缀参数，格式为 --ar x:y，如 --ar 3:4 或 --ar 9:16。如果在文本提示词的后面不输入 --ar，那么默认出图比例为 1 : 1，即图像是正方形的。正方形图像具有相等的宽度和高度，如 1000px×1000px 或 1500px×1500px 等。

　　--ar 后面的数值必须使用整数。例如，可以是 139：100 ，而不能是 1.39：1。某些图像内容只适用某些固定的比例，否则图像将会被裁切。放大图像时，某些图的宽高比可能会略有变化。

·最大宽高比例

　　Midjourney 不同版本支持生成的图片有着不同的最大宽高比，具体如下。生成的图片效果如图 2-82 所示。

　　V5 版本：支持任意比例关系。

　　V4 版本：宽高比最大支持 1：3，如 1：3 就无法支持。

　　V3 版本：宽高比最大支持 2：5。

　　Niji 版本：宽高比最大支持 1：2。

> 注：大于 2：1 的宽高比是实验性的，可能会产生不可预测的效果。

图 2-82

常见的图像比例如下表所示。

常见的图像比例				
--ar 1:1	--ar 3:4	--ar 5:4	--ar 3:2	--ar 16:9
默认宽高比	小红书App适用的最佳比例	常见的打印比例	常见单反相机拍摄出的照片比例	常见的屏幕比例

2.--q

--q（quality）指的是生成图像的质量。--q 值越大，生成的图像质量越高，花费的时间就越长。质量的改变不会影响图像的分辨率，只会影响图像的品质。

--q 参数值为 0.25~5，--q 值为 0.25 时出图速度最快，质量最差；--q 值为 5 时出图速度最慢，但质量最好。需要注意，对于 V4 和 V5 版本，参数值仅能从 0.25 变为 1，因此除非需要较低质量的图像，否则请跳过此参数。

注：--q 值默认为 1；--q 参数适用于 V1、V2、V3、V4、V5 和 Niji 版本。

·质量设置

--q 值越高不一定越好，有时较低的 --q 值可以产生更好的效果。较低的 --q 值可能适用于一些抽象的外观。较高的 --q 值可以将很多细节复杂的图像表现得惟妙惟肖，如雕花、建筑等。总之，选择与希望创建的图像类型最匹配的设置是最好的。

·提示示例

在 V5 版本模式下，--q .25 和 --q 2 的对比效果如图 2-83 和图 2-84 所示。

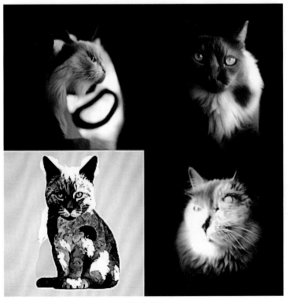

图 2-83

图 2-84

· 如何使用 --q 参数

将 --q 值放在文本提示词的后方即可, 如图 2-85 所示。

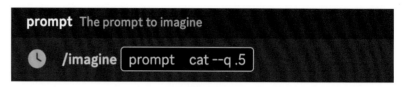

图 2-85

3.--s

--s 指的是风格化(--stylize), 可以影响默认的 Midjourney 美学风格, 数值为 0~1000, 默认值为 100; 数值不同, 美学风格不同。数值越低, 与文本提示词的内容就越匹配, 但是艺术性不强; 数值越高, 与文本提示词的内容关联性就越弱, 但非常具有艺术性。不同的 Midjourney 版本模型具有不同的风格化范围, 具体如下。

V1/V2: 不支持。

V3: 参数值为 625 时处于关闭状态, 默认值为 2500, 最大值为 60000。

V4/V5/Niji: 参数值为 0 时处于关闭状态, 默认值为 100, 最大值为 1000。

在 V5 版本模式下, --s 0、--s 500 和 --s 1000 的风格效果对比如图 2-86~ 图 2-88 所示。

图 2-86

图 2-87

图 2-88

· 如何使用 --s 参数

（1）将 --s 值放在文本提示词的后方即可。

图 2-89

（2）使用设置命令: 输入 /settings 并按 Enter 键, 单击 Style low、Style med、Style high、Style very high 中的任意一个按钮即可。

4.--v

Midjourney 会定期发布新模型版本，以提高作品质量和工作效率。不同的模型擅长处理不同类型的图像，出图时可以在文本提示词后面加 --v 参数，或使用 /settings 指令后，选择自己需要的模型版本。

> 注：--v 版本为 V1、V2、V3、V4 、V5、V5.1。

为了方便理解不同模型版本的特点，下面用同样的一组关键词（Vibrant tulip flower sea，生机勃勃的郁金香花海）来生成图像。

·--v 5.1 模型版本

此模型版本可以用更简单的文本描述生成美学风格更强的作品，如图 2-90 所示。它还具有较高的一致性，擅长准确解释提示内容，提高了图像的清晰度，并支持其他多项高级功能。

·RAW Mode

使用 RAW Mode 参数后生成的图像减弱了 Midjourney 默认的美学效果，非常适合想要控制生成的图像或想要生成摄影图像的用户，如图 2-91 所示。

图 2-90　　　　　　　　　　　　　　图 2-91

> 注：使用 RAW Mode 参数，需要将 --style raw 值放在文本描述的后方，如图 2-92 所示；或者使用设置命令，输入 /settings 后选择 RAW Mode 即可开启。

通过对比可以发现，使用 V5.1 生成的图像效果具有一定的艺术感与美感，而加了 RAW Mode 参数后，生成的图像更接近文本描述，且更接近真实照片。

图 2-92

--v 5 模型版本

此版本比默认的 5.1 版本能产生更好的摄影效果，并且生成的图像与文本提示非常匹配，如图 2-93 所示。不过，这个版本一般需要较长的文本提示才能生成满意的效果。

--v 4 模型版本

这个版本是 Midjourney 在 2022 年 11 月至 2023 年 5 月采用的默认模型版本。该版本具有全新的代码库和全新的架构，由 Midjourney 设计并在新的 Midjourney AI 超级集群上进行训练。与之前的模型相比，这个模型版本对生物、地点和物体的理解能力有所增强，用这个版本生成的效果如图 2-94 所示。

图 2-93

图 2-94

· 以前的模型版本

可以使用 --v 或 /settings 指令来选择更早的模型版本，不同的模型版本擅长处理不同类型的图像，V1、V2 和 V3 模型版本生成的图像效果如图 2-95 ~ 图 2-97 所示。

图 2-95

图 2-96

图 2-97

通过对比可以看出，V5.1 和 V5 生成的图像效果较好，其中 V5.1 版本开启 RAW Mode 参数后，生成的图像更接近摄影作品。而 V4 版本生成的效果较为一般，V1、V2 和 V3 版本可以直接忽略。

--niji 模型版本

niji 模型版本是 Midjourney 和 Spellbrush 合作研发出来的，可以生成动漫和插画风格的图像，它在动态和动作镜头以及以角色为中心的构图方面表现较为出色。目前有两个版本，即 niji 4 和 niji 5。在关键词末尾添加 --niji 就可以生成非常精美的二次元动漫风格的图像。使用 niji 4 和 niji 5 生成的效果如图 2-98 和图 2-99 所示。

图 2-98

图 2-99

· niji 样式参数

在 niji 模型版本中还有很多不同的细分风格，分别是可爱风格（--style cute）、风景优美的风格（--style scenic）、原始风格（--style original）、具有表现力的风格（--style expressive），生成的效果如图 2-100～图 2-103 所示。

图 2-100

图 2-101

图 2-102

图 2-103

·如何切换模型

（1）直接将 --v 1、--v 2、--v 3、--v 4、--v 5、--v 5.1、--niji4、--niji5 等添加到提示词的末尾即可。

（2）使用设置命令：键入 /settings 并在设置面板中选择需要的版本。

图 2-104

图 2-105

5.--c

　　--c 值为 0~100。--c 值默认为 0；--c 值越小，图像效果越符合描述指令；--c 值越大，图像越具有意想不到的构图或艺术效果。

　　使用较低的 --c 值或不指定值，生成的图像通常略有不同，如图 2-106 所示（案例数值：--c 10）。

图 2-106

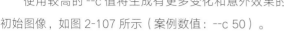

使用较高的 --c 值将生成有更多变化和意外效果的初始图像，如图 2-107 所示（案例数值：--c 50）。

使用极高的 --c 值生成的初始图像具有意想不到的构图或艺术效果，如图 2-108 所示（案例数值：--c 100）。

图 2-107

图 2-108

· 使用 --chaos 或 --c 参数

将 --chaos \<value\> 或 --c \<value\> 添加到提示词的末尾即可，如图 2-109 所示。

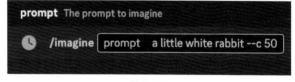

图 2-109

6.--seed

Seed 值是每个图像随机生成的，可以使用 --seed 或 --sameseed 参数指定，使用相同的种子编号和提示词将生成相似的图像。

· 如何查询 Seed 值

可在 Discord 中单击图像右上方的表情图标，然后单击信封图标，私信里会收到四宫格图像的 Seed 值和 Job ID（此处可参考"2.1.5 /show 指令"小节），如图 2-110~ 图 2-113 所示。

图 2-110

图 2-111

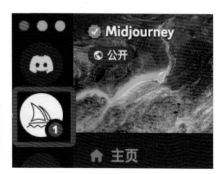

图 2-112

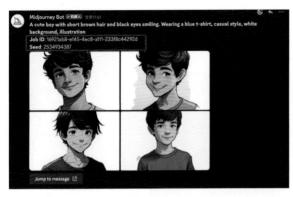

图 2-113

> 注：
>
> Seed 值是 0~4294967295 中的一个整数。
>
> Seed 值仅影响初始四宫格图像，升档后（单击 U/V 按钮以后）的 Seed 值无效。
>
> --sameseed 参数适用范围：在 V1、V2、V3、MJ Test 和 MJ Test Photo 版本模式下，将生成具有相似构图、颜色和细节的图像。
>
> 使用 V4、V5 和 Niji 的相同种子值，会生成几乎相同的图像。

· 如何获取过去图像的种子编号

如果想获取过去图像的种子编号，可以复制 Job ID 并使用 /show <Job ID #> 具有该 ID 的指令来恢复作业。然后使用信封表情符号对新生成的作业做出反应。Midjourney 每次生成的图像都不一样，如果想生成一个系列的图像，如同一个小男孩的不同表情，那就需要使用 Seed 值修改生成好的图像参数，具体操作如下。

（1）生成一组小男孩图像，文字描述如下。找到它的 Seed 值并复制。

a cute boy with short brown hair and black eyes crying. wearing a blue t-shirt, casual style, white background, illustration

一个可爱的男孩，棕色的短发，黑色的眼睛，哭泣。穿着蓝色的 T 恤，休闲风格，白色的背景，插画

图 2-114

（2）重新输入新的文字描述，可以将"哭泣"改为"愤怒"，在末尾加上 --seed 值。

a cute boy with short brown hair, black eyes, angry. wearing a blue t-shirt, casual style, white background, illustration

一个可爱的男孩，棕色的短发，黑色的眼睛，愤怒。穿着蓝色的 T 恤，休闲风格，白色的背景，插画

图 2-115

·用 Seed 值保留构图，衍生出不同场景

superman and batman, standing in a bar, chatting with glasses of wine, at night, very realistic

超人和蝙蝠侠，站在酒吧里，拿着酒杯聊天，在晚上，非常逼真

图 2-116

superman and beauty, standing in the bar, chatting with wine glasses, at night, very realistic

超人和美女，站在酒吧里，拿着酒杯聊天，在晚上，非常逼真

图 2-117

　　按照上述思路，我们除了可以更改人物表情、场景，还可以更改作品中的物品、天气、灯光等。Seed 值对画面的影响有限，且受时间、随机性等影响，单独使用 Seed 值去微调画面是比较困难的。在背景简单，事物描述具体、清晰，画面风格明显，且构图简单的图像上使用这种方法比较可行。

·如何使用种子编号

　　使用 --seed 或 --sameseed 参数，将 --seed <value> 或 --sameseed <value> 添加到提示词的末尾即可，如图 2-118 所示。

图 2-118

7.--video

　　--video 是用来创建一幅图像生成过程的视频的指令。该指令只适用于 V1~V3、MJ Test 和 MJ Test Photo 版本。图像生成后，单击信封图标，Midjourney Bot 会发来图形和视频，以供下载。如果图像质量一般，那么视频时长较短；如果图像质量很高，那么视频时长就会很长。

> 注：
> --video 仅适用于初始图像，不适用于高档图像（单击 U 和 V 按钮以后）。

· 如何获取视频链接

（1）输入文本提示词后，将 --video 添加到提示词的末尾。

图 2-119

（2）图像生成后，单击表情图标。

图 2-120

（3）单击信封图标。

图 2-121

（4）Midjourney Bot 机器人会将视频链接发送到直接消息中。

图 2-122

（5）单击链接，可在浏览器中查看并下载视频。

图 2-123

· 如何使用 --video 参数

将 --video 添加到文本提示词的末尾即可，如图 2-124 所示。

图 2-124

8.--tile

--tile 可以用来创建一幅无缝纹理的图像。可以将该图像平铺为纺织物纹理、壁纸和无缝贴图等，如图 2-125 和图 2-126 所示。--tile 适用于 V1、V2、V3、V5 版本及 MJ Test、MJ Test Photo 测试版本。

图 2-125

图 2-126

· 如何使用 --tile 参数

将 --tile 添加到文本提示词的末尾即可，如图 2-127 所示。

图 2-127

9.--stop

通常在生成一幅图像时，都会有一个生成进度：0~100%。如果想生成一些比较抽象的画面，可以通过中途停止图像生成来达到这种效果。这时就要用到 --stop 参数。使用不同参数值生成的效果不同，--stop 值为 10、50、100 时的出图效果如图 2-128~图 2-130 所示。

注：--stop 接受值为 10~100，默认值为 100。

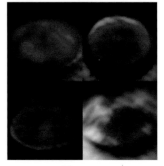

图 2-128

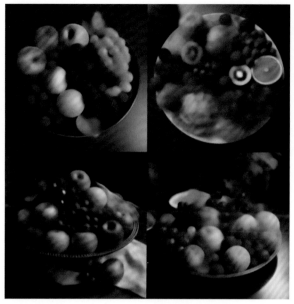

图 2-129

图 2-130

· 如何使用 --stop 参数

将 --stop+ 数值放在文本提示词的末尾即可，如图 2-131 所示。

图 2-131

2.1.8 隐身模式

Midjourney 是一个默认开放的社区，所有图像都可以在 Midjourney 官网上和私人 Discord 服务器中生成。而 Pro Plan 的订阅用户可以访问隐身模式，以防止他们生成的图像在 Midjourney 网站上被其他人看到；也可以在隐身模式（/stealth）和公开模式（/public）之间切换。

> 注：隐身模式只能防止其他人在 Midjourney 官网上查看你的图像；在公共频道中即使开启隐身模式，生成的图像别人也能看到，所以为防止其他人看到，请在私人 Discord 服务器中生成图像。

2.2 Midjourney出图技巧

掌握 Midjourney 的基本语法结构，学会文本的描述技巧，更好地理解语法、句子结构、单词及指令参数，可以得到更符合要求的创意图片。

2.2.1 基本语法结构

1. 初级指令

初级指令可以是一个单词、短语或表情符号等。例如，可以输入 a cat（一只猫），如图 2-132 所示。

图 2-132

2. 进阶指令

进阶指令是指在初级指令后面添加的后缀参数，用来调整图像的宽高比和出图的版本等。例如，要想获得一幅宽高比为 3∶4 且用 V5 版本生成的图，就可以在 a cat 后面加上 --ar 3:4 和 --v 5，如图 2-133 所示。

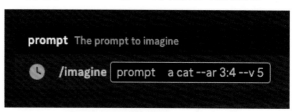

图 2-133

3. 高级指令

高级指令可以包括一幅或多幅图像的URL（上传的图片链接）、一个或多个描述短语，以及一个或多个后缀参数，如图 2-134 所示。

图 2-134

· 图像提示词

又称"垫图"或"喂图"。可以将图像的 URL 添加到提示词中，以影响最终生成图像的样式和内容。URL 要始终位于文本提示词的前面，输入文本提示词前一定要按 Space 键。

· 文本提示词

编写要生成图像的文本描述，精心编写的提示词有助于生成令人惊叹的图像。

· 后缀参数

后缀参数可以用来设置宽高比、出图的模型版本、放大器等。后缀参数必须放在提示词的末尾。

2.2.2 文本描述概述

因为 Midjourney Bot 不会像人类那样理解语法、句子结构或单词，所以单词的选择很重要，更少的单词意味着每个词都有更大的影响力。描述的语句要尽可能地清晰，含糊其词的描述往往无法获得所需的具体细节。例如，"3只猫"比"一些猫"更为具体。可以使用逗号、括号和连字符来组织语言，以形成完整的描述语句。文本描述的内容主要包括主题、媒介、环境、照明、颜色、情绪和构图等。

文本描述的主要内容						
主题	媒介	环境	照明	颜色	情绪	构图
人	照片	室内	柔和	充满活力	稳重	人像
动物	绘画	室外	环境	柔和	平静	爆头
地点	插图	月球上	阴天	明亮	喧闹	特写
物体	雕塑	水下	霓虹灯	单色	精力充沛	鸟瞰图
……	涂鸦	翡翠城	工作室灯	彩色	……	……
	挂毯	……	……	黑白		
	……			……		

输入不同的文本描述可以获得不同的图像，如图 2-135 和图 2-136 所示。

图 2-135

20 世纪 30 年代　20 世纪 40 年代　20 世纪 50 年代　20 世纪 60 年代　20 世纪 70 年代

坚定的　快乐的　困倦的　生气的　害羞的

苔原　盐滩　丛林　沙漠　山

图 2-136

2.2.3 出图技巧

　　了解了各项指令参数及基本的语法结构之后，新手初次出图时也许还是很难生成理想的效果。这是因为还没有真正掌握出图的技巧。下面具体讲解出图技巧。

　　一幅完整的图像需要具备以下几方面的内容：主题内容 + 细节描述 + 风格类型 + 指令参数。

主题内容 + 细节描述 + 风格类型 + 指令参数 = 完整作品	
主题内容	人物、地点、动作……
细节描述	质感、光线、构图……
风格类型	艺术家风格、流派、属性……
指令参数	渲染器、品质、尺寸……

　　这里用一个 IP 形象来举例说明。从主题内容切入，首先笔者想要的是一个小姑娘的形象：穿着白色的裙子，在一片雨林中奔跑，像一位探险家一样。相信这时大家的脑海中应该有画面了，再稍加润色，即可得到一组文本描述："一个非常可爱的女孩，穿着白色裙子，在雨林中奔跑，神秘的冒险"。将描述内容翻译成英文并填入指令框中，会得到图 2-137 所示的一组图像。

图 2-137

关键词组合

a very cute girl, wearing a white dress, running in the rainforest, a mysterious adventure.

一个非常可爱的女孩，穿着白色裙子，在雨林中奔跑，神秘的冒险

　　生成的图像看上去还不错,但并不是笔者想要的IP形象,所以需要继续添加细节描述。这里的细节描述包括细节、质感、光线、构图等。笔者想在画面中添加一些萤火虫、小鸟、各种植物,以及各种特殊的光线效果,再加上之前的主题描述内容,即可得到一组文本描述。将文本描述翻译成英文并填入指令框中,可以得到图 2-138 所示的一组图像。

图 2-138

关键词组合

a very cute girl, wearing a white dress, running in the rainforest, a mysterious adventure. fireflies, birds, plants, intricate details, animation lighting, uplight

一个非常可爱的女孩,穿着白色裙子,在雨林中奔跑,神秘的冒险。萤火虫,鸟,植物,复杂的细节,动画照明,亮光可以发现图像内容变得更加丰富了，但是与 IP 形象还有一定的差距，可以再加上一些 IP 形象的风格描述，如 Popmart 盲盒、黏土材料等。将补充后的文本描述翻译成英文并填入指令框中，会得到图 2-139 所示的一组图像。

图 2-139

关键词组合

a very cute girl, wearing a white dress, running in the rainforest, a mysterious adventure. fireflies, birds, plants, intricate details, animation lighting，uplight, popmart blind box, clay material, pixar trend, 3d art, c4d

一个非常可爱的女孩，穿着白色裙子，在雨林中奔跑，神秘的冒险。萤火虫，鸟，植物，复杂的细节，动画照明，亮光，泡泡玛特盲盒，黏土材料，皮克斯趋势，3D 艺术，C4D

　　到这一步，IP 形象的效果就基本出来了，但感觉还是不够精致，欠缺一些质感，画面的比例也不对。所以还要添加一些指令参数，如渲染器、比例、风格化程度等。这样一组精致的 IP 形象就生成了，如图 2-140 所示。

关键词组合

a very cute girl, wearing a white dress, running in the rainforest, a mysterious adventure. fireflies, birds, plants, intricate details, animation lighting, uplight, popmart blind box, clay material, pixar trend, 3d art, c4d. octane rendering, hd, --ar 3:4 --v 5 --s 750

一个非常可爱的女孩，穿着白色裙子，在雨林中奔跑，神秘的冒险。萤火虫，鸟，植物，复杂的细节，动画照明，亮光，泡泡玛特盲盒，黏土材料，皮克斯趋势，3D 艺术，C4D, Octane 渲染，高清，--ar 3:4 --v 5 --s 750

图 2-140

　　以上讲解的是生成一组完整作品的大致思路和技巧，只要按照这个思路和技巧操作，基本都可以生成自己想要的各种作品。当然，大家还需要掌握多种艺术风格、设计名词、光线效果、构图视角等。拓宽自己的眼界，提升审美能力，多思考，多尝试，有利于生成自己所需的图像。

VISUAL
PERFORMANCE

视觉表现

3.1 绘画类

关键词组合

line draft style, dog with sunglasses and jacket, riding a motorcycle, no background, very detailed --ar 3:4--q 2

线稿风格，狗戴着太阳镜，穿着夹克，骑着摩托车，没有背景，非常详细 --ar 3：4 --q 2

关键词组合

line draft style, girl, plants, no background, very detailed --ar 3:4 --q 2

线稿风格，女孩，植物，无背景，非常详细 --ar 3：4 --q 2

关键词组合

line draft style, portrait of a man, no background, manuscript, black and white, very detailed --ar 3:4 --q 2

线稿风格，男子肖像，无背景，手稿，黑白，非常详细 --ar 3：4 --q 2

关键词组合

detailed pencil art of a full body portrait: china perfect female face: large soulful eyes: full lips: with eyeglasses: in the style of sciamano240, anime --q 2

详细的铅笔艺术的全身肖像：中国完美的女性面孔：深情的大眼睛：丰满的嘴唇：戴着眼镜：Sciamano240 的风格，动画 --q 2

关键词组合

an old man was picking up garbage in the street, line draft style, portrait of a man, no background, manuscript, black and white, very detailed --ar 3:4 --q 2

一位老人在街上捡拾垃圾，线稿风格，人物肖像，无背景，手稿，黑白，非常详细 --ar 3：4 --q 2

关键词组合

shanghai busy old street scene, christmas, snow falling heavily, soft watercolour style, ink painting, flat design aesthetic, perfect composition, masterpiece --ar 3:4 --q 2

上海繁华老街景，圣诞节，大雪纷飞，柔和水彩风格，水墨画，平面设计美学，完美构图，杰作 --ar 3：4 --q 2

关键词组合

soft watercolor style, coby whitmore ink drawings, graphic design aesthetics, christmas, renaissance street scenes, heavy snowfall --ar 3:4 --q 2

柔和的水彩风格，Coby Whitmore 水墨画，平面设计美学，圣诞节，文艺复兴时期的街景，大雪 --ar 3：4 --q 2

关键词组合

girl with long hair, portrait, 8 flowers, watercolor, 4k, hd, original --ar 3:4 --q 2

长头发的女孩，人像，8 朵花，水彩，4K，高清，原创 --ar 3：4 --q 2

关键词组合

a sika deer with flowers on its antlers, close-up, watercolor, flora --ar 3:4 --q 2

鹿角上有花的梅花鹿，特写，水彩，植物群 --ar 3：4 --q 2

关键词组合

the little girl should run with a dog in the countryside, by hayao miyazaki, studio ghibli, watercolor, very detailed --ar 3:4 --q 2

小女孩在乡下和狗一起奔跑，宫崎骏，吉卜力工作室，水彩，非常详细 --ar 3：4 --q 2

关键词组合

watercolor style, streets of paris after the rain, eiffel tower, refreshing images, people coming and going --ar 3:4

水彩风格，雨后的巴黎街头，埃菲尔铁塔，清新的画面，人来人往 --ar 3：4

关键词组合

the ink mecha characters fuse with the landscape all over their bodies, freehand, dancing rivers and lakes, swordsmen, black and white paintings, ink art --ar 3:4 --s 250 --v 5 --q 2

水墨画人物，全身与山水融合，写意，舞动江湖，剑客，黑白画，水墨艺术 --ar 3：4 --s 250 --v 5 --q 2

关键词组合

Riverside Scene at Qingming Festival, ink painting --ar 16:9 --v 5 --q 2

《清明上河图》，水墨画 --ar 16：9 --v 5 --q 2

关键词组合

chinese painting, ink painting, award-winning work. a huge tall pale vine with white flowers grows, surrounded by white silk, bird, with a morandi color scheme, aesthetic and symmetrical composition. by qi baishi --ar 16:9 --q 2 --upbeta --v 5

中国画，水墨画，获奖作品。一棵巨大的高大的淡色藤蔓上生长着白色的花朵，周围是白色的丝绸，鸟儿，有莫兰迪的配色，唯美而对称的构图。齐白石 --ar 16：9 --q 2 --upbeta --v 5

关键词组合

chinese super beautiful illustrations, tradition chinese ink painting, stunning natural landscape, ink illustration, ink wash painting style, beijing, spring city, exquisite details --ar 3:4 --s 250 --v 5 --q 2

超级美丽的中国插图，传统中国水墨画，惊人的自然景观，水墨插图，水墨画风格，北京，春城，精致的细节 --ar 3：4 --s 250 --v 5 --q 2

关键词组合

a young man holding a long sword, from a chinese martial arts novel, with long black hair fluttering in the wind tall, with long arms and legs, wearing armor looks suave. ink painting --q 2 --s 750 --v 5 --ar 3:4

一个手持长剑的年轻人，来自中国的武侠小说，黑色的长发在风中飘扬，身材高大，手脚修长，穿着盔甲，看起来很有风度。水墨画 --q 2 --s 750 --v 5 --ar 3：4

关键词组合

a magical beauty with flowing long hair, wearing a magic uniform, especially beautiful, fighting pose, japanese animation, makoto shinkai style, aesthetic, exquisite, smooth lines, beautiful light shadow and color --ar 3:4 --q 2 --v 5

留着飘逸长发的魔法美女，穿着魔法制服，特别漂亮，战斗姿势，日本动画，新海诚风格，美学，精致，线条流畅，美丽的光影和色彩 --ar 3：4 --q 2 --v 5

关键词组合

boy, 8k, cute, anime, acgn, illustration, beautiful light shadow and color --q 2 --v 4

男孩，8k，可爱，动画，ACGN，插图，美丽的光影和色彩 --q 2 --v 4

关键词组合

girl, 8k, cute, anime, acgn, illustration, beautiful light shadow and color --q 2 --v 4

女孩，8k，可爱，动画，ACGN，插图，美丽的光影和色彩 --q 2 --v 4

关键词组合

magician, pretty girl, 8k, cute, anime, acgn, beautiful light and color --q 2 --niji 5

魔术师，美女，8K，可爱，动画，ACGN，美丽的光影和色彩 --q 2 --niji 5

关键词组合

vampire, handsome boy, 8k, cute, anime, acgn, beautiful light and color --q 2 --niji 5

吸血鬼，帅哥，8K，可爱，动画，ACGN，美丽的光影和色彩 --q 2 --niji 5

关键词组合

boys and girls, high school students, coming home from school, walking on the road, sunset, cherry blossom, healing department, aesthetic, hayao miyazaki style, ghibli --ar 3:4 --q 2 --v 4

男孩和女孩，高中生，放学回家，走在路上，日落，樱花，美学，治愈系，宫崎骏风格，吉卜力 --ar 3：4 --q 2 --v 4

关键词组合

japanese animation, makoto shinkai style, a girl with long hair spreads her arms, very beautiful, with exquisite facial features, looking up at the sky. expressed a hug gesture to the camera. there are many clouds in the evening sky, and colorful pops across the sky. birds, feathers, the picture is in the middle ground, the style is similar to xin haicheng's works. aesthetic, simple and clean pictures, cherry blossom --ar 3:4 --q 2 --v 5

日本动画，新海诚风格，一个长发女孩张开双臂，非常漂亮，五官精致，仰望天空。对着镜头表现出一个拥抱的姿态。傍晚的天空中有许多云彩，五颜六色的流行元素横跨天空。鸟儿，羽毛，画面处于中间位置，风格与新海诚的作品相似。美学，画面简单干净，樱花 --ar 3：4 --q 2 --v 5

关键词组合

daniel f.gerhartz style, oil painting, contrast between cold and warm, a real forest town made, fine wool grain texture surface --v 5 --ar 3:4 --q 2

丹尼尔·格哈茨风格，油画，冷与暖的对比，一个真正的森林小镇，精细的羊毛纹理表面 --v 5 --ar 3：4 --q 2

关键词组合

Oil painting, field at dusk, farmers are harvesting wheat in the field, some children are playing beside, Vincent van Gogh style --ar 16:9 --q 2

油画，黄昏的田野，农民在田里收割小麦，一些孩子在旁边玩耍，文森特·凡·高风格 --ar 16：9 --q 2

关键词组合

art by raffaello sanzio, oil painting,abstract oil paintings, people coming and going, the hustle and bustle of the city, busy figures, dusk after rain,serene ambience --ar 16:9 --q 2

拉斐尔·圣齐奥的艺术作品，油画，抽象油画，人来人往，城市的喧嚣，忙碌的身影，雨后的黄昏，宁静的氛围 --ar 16：9 --q 2

关键词组合

oil painting, imposing church, divine light, priest and people praying in the square, plants, style of leonardo da vinci --ar 3:4 --q 2

油画，雄伟的教堂，神圣的光芒，牧师和广场上祈祷的人们，植物，达·芬奇的风格 --ar 3：4 --q 2

关键词组合

early morning sea level, the sun rises, mist covers the sea, a woman walks by the sea, oil painting, monet style --ar 3:4

清晨的海平面，太阳升起，薄雾笼罩着大海，一个女性在海边散步，油画，莫奈风格 --ar 3：4

关键词组合

a beautiful girl, morandi tones, oil painting, refreshing images, in style of abbott fuller graves --ar 3:4 --v 5 --q 2

一个美丽的女孩，莫兰迪的色调，油画，令人耳目一新的图像，阿博特・富勒・格雷夫斯的风格 --ar 3：4 --v 5 --q 2

关键词组合

oil painting, blue flowers in the field, soft and dreamy atmosphere, extensive use of palette knife, joyful natural scenery, sky blue light green light blue, impressionist landscape painting style --ar 3:4 --s 750 --v 5

油画，田野里的蓝色花朵，柔和梦幻的氛围，大量使用调色刀，欢乐的自然风光，天蓝色 浅绿色 浅蓝色，印象派风景画风格 --ar 3：4 --s 750 --v 5

关键词组合

female figure in the center of the frame, classical ornament, roses, leaves, buds, feathers, printmaking, print, old book illustrations, relief printing, intaglio printing, etching, line engraving, woodblock printing, wood engraving, steel engraving, copper engraving, vivid, contrasting, nature vibes, light, sharp --ar 3:4 --q 2

框架中央的女性形象，古典装饰，玫瑰，树叶，花蕾，羽毛，版画复制，版画，旧书插图，凸版印刷，凹版印刷，蚀刻，线雕，木版印刷，木版画，钢版画，铜版画，生动，对比鲜明，自然氛围，轻盈，锐利 --ar 3：4 --q 2

关键词组合

highly detailed and intricate illustration for the wilderness, printmaking, print, old book illustrations, relief printing, intaglio printing, etching, line engraving, woodblock printing, wood engraving, steel engraving, copper engraving, vivid, contrasting, nature vibes, light, sharp --ar 3:4 --q 2

非常详细和复杂的荒野插图，版画复制，版画，旧书插图，凸版印刷，凹版印刷，蚀刻，线条雕刻，木版印刷，木版画，钢版画，铜版画，生动，对比鲜明，自然氛围，轻盈，锐利 --ar 3：4 --q 2

关键词组合

illustrations, printmaking, print, old book illustrations, relief printing, intaglio printing, etching, line engraving, woodblock printing, wood engraving, steel engraving, copper engraving --ar 3:4 --q 2

插图，版画复制，版画，旧书插图，凸版印刷，凹版印刷，蚀刻，线雕，木版印刷，木版画，钢版画，铜版画 --ar 3：4 --q 2

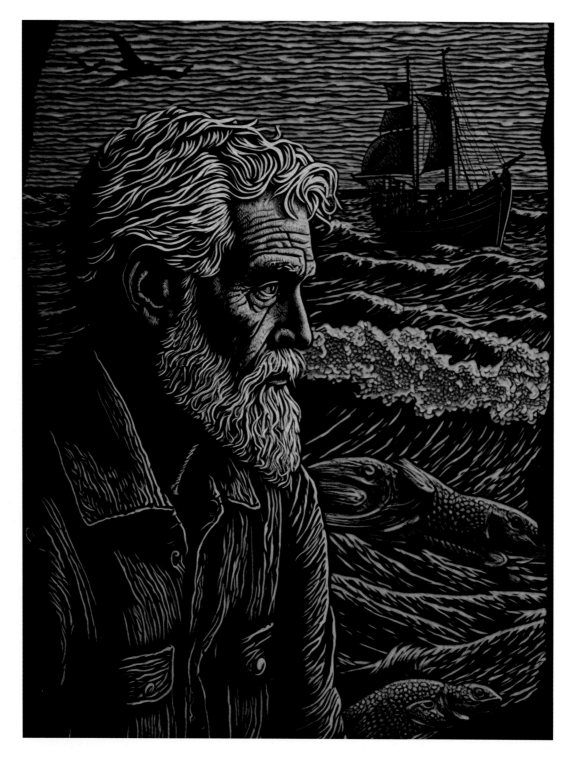

关键词组合

old man and the sea woodcut, 3d model, intricate,printmaking, print, old book illustrations, relief printing, intaglio printing, etching, line engraving, woodblock printing, wood engraving, steel engraving, copper engraving, vivid, contrasting --ar 3:4 --q 2

老人与海木刻，3D 模型，复杂，版画复制，版画，旧书插图，凸版印刷，凹版印刷，蚀刻，线雕，木版印刷，木版画，钢版画，铜版画，生动，对比鲜明 --ar 3：4 --q 2

关键词组合

ukiyo-e, samurai vs kendo, pagoda, cherry blossoms, woodblock print --ar 3:4

浮世绘，武士对剑道，宝塔，樱花，木版画 --ar 3：4

关键词组合

Japanese art, ukiyo-e, sunsets after snow are so beautiful, I rode my horse through the countryside, highly detailed, 4K --q 2 --ar 16:9 --v 5

日本艺术，浮世绘，雪后的夕阳如此美丽，我骑着马穿过乡村，非常详细，4K--q 2 --ar 16：9 --v 5

关键词组合

ukiyo-e, kanagawa, okinawa, a fisherman catches fish in the waves --ar 16:9 --v 5

浮世绘，神奈川，冲绳，一个渔夫在海浪中捕鱼 --ar 16：9 --v 5

关键词组合

ukiyo-e style, woman in japan --ar 3:4 --q 2 --v 5

浮世绘风格，日本女人 --ar 3：4 --q 2 --v 5

关键词组合

snow appreciation in japan, ukiyoe, by katsushika hokusai --q 2 --ar 3:4 --v 5

在日本赏雪，浮世绘，葛饰北斋所作 --q 2 --ar 3：4 --v 5

关键词组合

female asian woman with blue dress and birds in garden, in the style of silk painting, dark white and light red, uemura shoen, historical reproductions, elegant, emotive faces, mural painting --ar 51:91 --s 250 --v 5

公园中穿蓝裙的亚洲女子与鸟，绢画风格，深白色和浅红色，上村松园，历史复制品，优雅，表情丰富的脸，壁画 --ar 51：91 --s 250 --v 5

关键词组合

a chinese painting with trees on it, light pink and dark brown, trompe l'oeil details, superb shadows, intricate scenery, twisted branches, hand scroll, high resolution, traditional scroll painting style --ar 1:4 --s 750 --v 5

一幅有树的中国画，浅粉色和深棕色，错觉细节，精湛的阴影，复杂的风景，扭曲的树枝，手卷，高分辨率，传统卷轴画风格 --ar 1：4 --s 750 --v 5

关键词组合

two birds standing on a tree branch, in the style of meticulous brushwork, flower and nature motifs, hasselblad h6d-400c, mural painting, beige, mingei, romanticized realism --ar 1:4 --s 250 --v 5

两只站在树枝上的鸟，工笔画风格，花卉和自然图案，哈苏 H6D-400C，壁画，米色，民艺，浪漫的现实主义 --ar 1：4 --s 250 --v 5

关键词组合

chinese painting of an asian flower in a flower, in the style of white and brown, delicate still-lifes, richly layered, porcelain, wallpaper, bold yet graceful --ar 1:4 --s 250 --v 5

中国花鸟画，白色和棕色的风格，精致的静物画，层次丰富，瓷器，墙纸，大胆而优雅 --ar 1：4 --s 250 --v 5

关键词组合

this painting depicts an oriental lady in a traditional dress, in daz3d style, delicate marks, influenced by ancient chinese art, dark green and yellow, anime style character design, delicate portrait, romantic depiction of historical events --ar 51:91 --s 250 --v 5

这幅画描绘了一位穿着传统服装的东方女士，采用Daz3D风格，精致的标记，受中国古代艺术的影响，深绿色和黄色，动漫风格的人物设计，精致的肖像，对历史事件的浪漫描绘 --ar 51：91 --s 250 --v 5

关键词组合

a woman is holding a golden bowl of food in front of an oriental painting, in the style of daz3d, xu beihong, elegant inking techniques, serene faces, 500–1000 ce, romantic illustrations, detailed facial features --ar 51:91 --s 250 --v 5

一个女人在东方绘画前拿着一个金碗，采用Daz3D的风格，徐悲鸿，优雅的水墨技术，宁静的面孔，公元500—1000年，浪漫的插图，详细的面部特征 --ar 51:91 --s 250 --v 5

关键词组合

angel statues, 3d printed, in the style of photorealistic renderings, françois boucher, mixes realistic and fantastical elements, 8k resolution, lush and detailed, salvator rosa, sergey marshennikov --ar 51:91 --s 250 --v 5

天使雕像，3D 打印，逼真的渲染风格，弗朗索瓦·布歇，混合现实和幻想的元素，8K 分辨率，丰富而细致，萨尔瓦托·罗萨，谢尔盖·马什尼科夫 --ar 51：91 --s 250 --v 5

关键词组合

a white statue with wings and candles, in the
style of vray tracing, james bullough, delicate
flora depictions, 32k uhd, jean-léon gérôme,
natalie shau, delicate materials --ar 23:50 --s 250
--v 5

带有翅膀和蜡烛的白色雕像，VRay 追踪的风
格，詹姆斯·布洛，精致的植物描写，32K
超高清，让·莱昂·杰罗姆，娜塔莉·肖，精
致的材料 --ar 23：50 --s 250 --v 5

关键词组合

a wooden figurine that looks like a buddha sitting, in the style of precisionist art, organic formations, hyper-realistic sculptures, restored and repurposed, swirling vortexes, naturalistic renditions, nikon af600 --v 5

一个看起来像佛祖坐着的木制雕像，精确主义艺术风格，有机的形式，超现实的雕塑，修复和再利用，漩涡，自然主义的演绎，尼康 AF600--v 5

关键词组合

a buddha carving with some carved branches, meticulous linework precision, erik jones, shigeru aoki, hurufiyya, use of precious materials, swirling vortexes --v 5

雕刻着树枝的佛像，细致的线条，埃里克·琼斯，青木茂，胡鲁菲亚，使用珍贵的材料，漩涡 --v 5

关键词组合

pink flamingo, the bird of the flower wonderland, in the style of rendered in cinema4d, rococo-inspired details, textural explorations, monochromatic color scheme, 32k uhd, hyper-realistic animal illustrations, fluid form --ar 2:3 --v 5

粉红色的火烈鸟，仙境之鸟，Cinema 4D 渲染风格，洛可可风格的细节，纹理的探索，单色方案，32K 超高清，超现实的动物插图，流体形式 --ar 2：3 --v 5

关键词组合

a wooden ship is floating around in a piece of flame, in the style of intricately detailed patterns, dreamscape portraiture, high contrast lighting, bent wood, stenciled iconography, photographically detailed portraitures, lightbox --v 5

一艘木船在一片火焰中漂浮，具有复杂的细节图案风格，梦境肖像，高对比度的照明，弯曲的木头，模板印刷的图像，照片上的详细肖像，灯箱 --v 5

关键词组合

owl in wood carved art, in the style of detailed landscapes, nikon af600, 32k uhd, detailed engraving, detailed brushstrokes, twisted branches --v 5

木雕艺术中的猫头鹰，详细的风景画风格，尼康 AF600，32K 超高清，细节雕刻，细节笔触，扭曲的树枝 --v 5

关键词组合

sculpture, silver, lively portraits, aggressive digital illustration style, organic art nouveau, 8k 3d, luxurious wall hangings, ultra hd graphics --v 5

雕塑，银色，活泼的肖像，激进的数字插画风格，有机新艺术风格，8K 3D，豪华的壁挂，超高清图像 --v 5

关键词组合

the metal sculpture has numerous balls on it, in the style of high detailed, biomorphic --v 5

金属雕塑上有许多球，具有细节丰富的风格、生物形态 --v 5

关键词组合

one of many paper art pieces i'm working on, in the style of northern china's terrain, eye-catching composition, colored cartoon style, dark cyan and light bronze, majestic romanticism, swiss style, dreamlike symbolism --ar 98:129 --v 5

我正在创作的许多纸艺作品之一，中国北方地形的风格，醒目的构图，彩色卡通风格，深青色和浅青铜色，雄伟的浪漫主义，瑞士风格，梦幻般的象征主义 --ar 98：129 --v 5

关键词组合

graffiti artwork of a man in front of him, in the style of sabattier filter, dark pink and dark amber, dark white and sky-blue, hd, arabesque, i can't believe how beautiful this is, voluminous mass --ar 3:4 --v 5

一个侧面人物的涂鸦艺术作品，萨巴蒂尔滤镜风格，深粉色和深琥珀色，深白色和天蓝色，高清，阿拉伯花纹，我不能相信这是多么美丽，体量很大 --ar 3：4 --v 5

关键词组合

graffiti artwork of a man in front of him, in the style of sabattier filter, dark pink and dark amber, dark white and sky-blue, hd, arabesque, i can't believe how beautiful this is, voluminous mass --ar 3:4 --niji 5

一个正面人物的涂鸦艺术作品，萨巴蒂尔过滤器的风格，深粉色和深琥珀色，深白色和天蓝色，高清，阿拉伯花纹，我不能相信这是多么美丽，体量很大 --ar 3：4 --niji 5

关键词组合

an artistic mural of an owl with a bird, in the style of 32k uhd, urban and edgy, dark gray, colorful installations, outdoor scenes, expansive geometry --ar 16:9 --niji 5

一幅猫头鹰与鸟的艺术壁画，风格为 32K 超高清，城市和前卫，深灰色，多彩的装置，户外场景，膨胀的几何图形 --ar 16：9 --niji 5

关键词组合

a mural showing a woman with butterflies and flowers, in the style of dark gray and aquamarine, anamorphic art, fish-eye lens, gravity-defying architecture, sea and coast painter --ar 16:9 --v 5

一幅壁画，展示了一位戴着蝴蝶和花朵的女人，采用深灰色和海蓝宝石风格，变形艺术，鱼眼镜头，反重力建筑，海洋和海岸画家 --ar 16：9 --v 5

关键词组合

a street mural has been painted on the side of a building, afrofuturism, stan lee, 32k uhd, gray and bronze --ar 16:9 --v 5

建筑物侧面绘制了街头壁画，非洲未来主义，斯坦·李，32K 超高清，灰色和青铜色 --ar 16：9 --v 5

关键词组合

a mural of a beautiful woman, painted in pink, in the style of dark azure and aquamarine, realistic anamorphic art, neo-concrete, naturalistic ocean waves, time-lapse photography, intricate architectures, detailed character illustrations --ar 16:9 --v 5

一个漂亮女人的壁画，用粉红色画的，具有深天蓝色和海蓝色的风格，现实的变形艺术，新混凝土，自然主义风格的海浪，延时摄影，复杂的建筑，详细的人物插图 --ar 16：9 --v 5

关键词组合

8-bit pixel art, ancient castles, studio ghibli, cinematic stills, hdr --v 5

8 位像素艺术，古代城堡，吉卜力工作室，电影剧照，高动态范围成像 --v 5

关键词组合

8-bit pixel art, forbidden city, studio ghibli, cinematic stills, hdr --v 5

8 位像素艺术，紫禁城，吉卜力工作室，电影剧照，高动态范围成像 --v 5

关键词组合

8-bit pixel art, wuhan yellow crane tower, studio ghibli, cinematic stills, hdr --v 5

8 位像素艺术，武汉黄鹤楼，吉卜力工作室，电影剧照，高动态范围成像 --v 5

关键词组合

16-bit pixel art, peasant harvest scene, studio ghibli, cinematic still, hdr --ar 4:3 --v 5

16 位像素艺术，农民丰收场景，吉卜力工作室，电影剧照，高动态范围成像 --ar 4：3 --v 5

关键词组合

8-bit pixel art, shanghai cityscape 2048, movie stills --v 5

8 位像素艺术，上海城市景观 2048，电影剧照 --v 5

关键词组合

8-bit pixel art, tianjin cityscape 2048, movie stills --v 5

8 位像素艺术，天津城市景观 2048，电影剧照 --v 5

关键词组合

16-bit pixel art, guangzhou cityscape 2048, movie stills --v 5 --s 250

16 位像素艺术，广州城市景观 2048，电影剧照 --v 5 --s 250

关键词组合

16-bit pixel art, shanghai cityscape 2048, movie stills --v 5

16 位像素艺术，上海城市景观 2048，电影剧照 --v 5

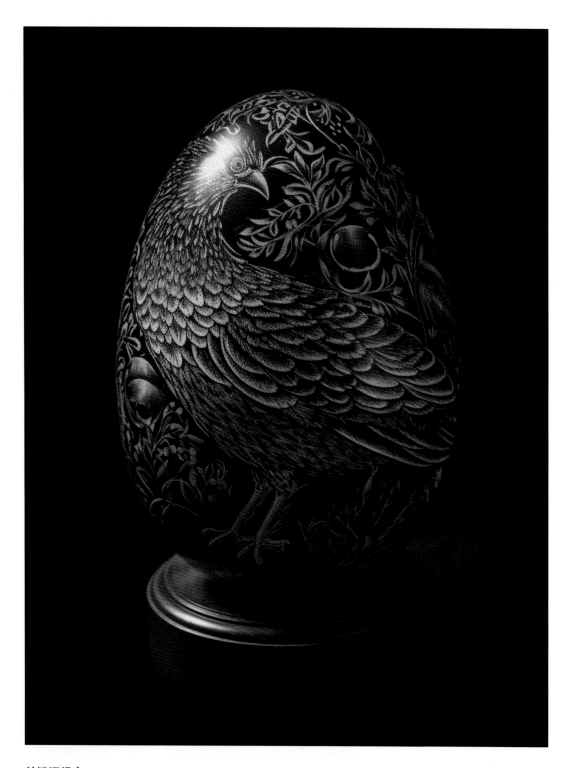

关键词组合

a wooden easter egg decorated with a picture of a bird on top, in the style of realism with fantasy elements, dark silver and dark gold, carved animal figures, intricate pen drawings, meticulous still lifes, black background, light beige and indigo --ar 3:4 --v 5

一个上面有鸟的图案的木制复活节彩蛋，带有幻想元素的现实主义风格，深银色和深金色，雕刻的动物形象，复杂的钢笔画，细致的静物画，黑色背景，浅米色和靛蓝色 --ar 3：4 --v 5

关键词组合

a wooden log with a sketch of mushrooms
and some flowers, in the style of mystical
creatures and landscapes, eco-friendly
craftsmanship, planar art, delicately detailed,
cottagecore, meticulous design --v 5

带有蘑菇和一些花朵草图的原木，神秘生
物和风景的风格，环保工艺，平面艺术，
精致细节，田园风，精心设计 --v 5

关键词组合

chinese painted flower in wood , in the style
of monochromatic masterpieces, lee bogle,
engraved ornaments, tondo, sepia tone, folk
art inspiration, serge marshennikov --v 5

中国画木花，单色杰作的风格，李·博格尔，
雕刻装饰品，通多，棕褐色调，民间艺术
灵感，塞尔日·马申尼科夫 --v 5

关键词组合

orchids with leaves on a wooden plate, in the style of meticulous linework precision, hyper-detailed illustrations, wood veneer mosaics, realistic anamorphic art, flower power, engraved line-work --v 5 --q 2

木盘上有叶子的兰花，以细致的线条风格，非常详细的插图，木面马赛克，逼真的变形艺术，花的力量，雕刻的线条作品 --v 5 --q 2

关键词组合

this image shows a monster in the sky, with fiery eyes, in the style of a digital fantasy landscape, intricately illustrated, emphasizing emotion over realism, flowing landscape, precision effect, detailed atmospheric portrait, serene face --ar 16:9 --q 2 --s 750 --v 5

这幅图片展示了天空中的一个怪物，有着炽热的眼睛，采用数字奇幻风景的风格，插图错综复杂，强调情感而不是现实主义，流动的风景，写实的效果，细致大气的肖像，平静的脸 --ar 16：9 -- q 2 --s 750 --v 5

关键词组合

dragon on top of mountain, high resolution, majestic, unobstructed ocean view, light golden gray, victorian genre painting, national geographic image --ar 16:9 --q 2 --s 750 --v 5

龙盘在山顶上，高分辨率，雄伟，一览无余的海景，淡金灰色，维多利亚风俗画，国家地理图片 --ar 16：9 --q 2 --s 750 --v 5

关键词组合

general in armor, tall and mighty, intricate background, exquisite details, original, rpg, 3a games, coohigh quality, complex details --ar 3:4 --v 5 --q 2

穿着铠甲的将军，高大威猛，背景错综复杂，细节精致，独创的，角色扮演游戏，3A 游戏，超高画质，细节复杂 --ar 3：4 --v 5--q 2

关键词组合

long-haired character standing on a red mountain with a sword in hand, dark gray and gold style, realistic ocean painting, hero masculinity, ue5, charming lighting --ar 3:4

长发人物手持着剑站在红山丘上，深灰色和金色风格，写实海洋画，英雄气概，UE5，迷人的灯光 --ar 3：4

关键词组合

image from anime series, armor holding great sword, aleksi briclot style, harsh palette knife work, madeira, violet and bronze, red and bronze, huge scale --ar 3:4

来自动漫系列的图像，盔甲拿着大剑，Aleksi Briclot 风格，粗糙的调色刀工作，马德拉，紫罗兰色和青铜色，红色和青铜色，巨大的规模 --ar 3：4

关键词组合

man in armor and trident, light amber and gray style, influenced by ancient chinese art, devil core, maid core, light gold and bronze, jungle core, unique character --ar 3:4

穿着盔甲且拿着三叉戟的男人，浅琥珀色和灰色的风格，受中国古代艺术的影响，魔鬼，女仆，浅金色和青铜色，丛林，独特的角色 --ar 3：4

关键词组合

the female who has the wings, and a golden moon and flowers, in the style of light red and dark azure, hyper-realistic water, 2d game art, colorful costumes, devilcore, li-core, silver and bronze --ar 3:4

长翅膀的女性，金色的月亮和花朵，浅红色和深蓝色风格，超写实的，2D游戏美术，多彩服饰，魔核，里核，银色和青铜色 --ar 3 : 4

关键词组合

a girl holding a sword in front of purple moons, in the style of light silver and dark azure, hyper-realistic illustrations, angura kei, light brown and azure, lively action poses, detailed costumes, dark blue and white --ar 3:4

一个女孩在紫色的月亮前持剑，浅银色和深蓝色风格，超写实插画，安古拉系，浅棕色和天蓝色，活泼的动作姿势，精致的服饰，深蓝色和白色 --ar 3：4

关键词组合

a fantasy figure in white and gold, in the style of ethereal symbolism, nightcore, animated gifs, beautiful women, wiccan, flowing fabrics, glassy translucence --ar 3:4

白色和金色的幻想人物，具有空灵象征主义的风格，夜，GIF 动画，美丽的女人，巫术崇拜者，流动的织物，玻璃般的半透明 --ar 3：4

关键词组合

a lady with white hair and a sword, in the style of dark gold and light beige, futuristic designs, white and brown, elaborate costumes, light bronze and black --ar 3:4

一位白发持剑的女士，深金色和浅米色风格，未来主义设计，白色和棕色，精致的服装，浅青铜色和黑色 --ar 3：4

关键词组合

a dark chinese structure with lights in it, in the style of detailed fantasy art, intricately textured, kevin hill, dark white and cyan, historical genre, fang lijun --ar 16:9

一个里面有灯光的深色的中式结构，详细的奇幻艺术风格，错综复杂的纹理，凯文·希尔，深白色和青色，历史类型，方力钧 --ar 16：9

关键词组合

the sky looms above the castle, the city sits under the bridge, futuristic urban style, realistic sculpture, deep white and gray, majestic port, skeuomorphic --ar 16:9

天空在城堡上空隐现，城市坐落在桥下，未来都市风格，写实雕塑，深白色和灰色，雄伟的港口，拟物化 --ar 16：9

关键词组合

game icons assets sheet. a collection of a variety of items of the kind, in the style of 2d game art, dark navy, realistic forms, rounded, enchanting, ironical, use of bright colors --v 5

游戏图标资产表。各种物品的集合，采用 2D 游戏艺术风格，深海军蓝，逼真的形式，圆形的，迷人，讽刺，使用鲜艳的色彩 --v 5

关键词组合

game icons assets sheet 5x5 of magic legendary, epic，games icons, rpg skills icons, honor of kings items icons, league of legends items icons, ability icon, flat pile gray background，octane rendering, 4k, raytracing --q 2

游戏图标资产表 5x5 魔法传奇，史诗，游戏图标，角色扮演游戏技能图标，《王者荣耀》物品图标，《英雄联盟》物品图标，能力图标，平堆灰色背景，Octane 渲染，4K，光线追踪 --q 2

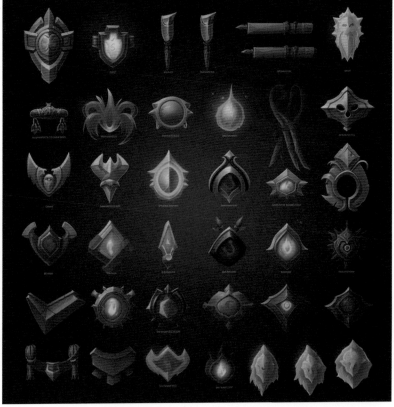

关键词组合

game icons assets sheet. a collection of many colorful icon designs, in the style of 2d game art, 32k uhd, detailed hyperrealism, indigo and bronze, stone, elaborate spacecrafts, dom qwek --v 5

游戏图标资产表。许多丰富多彩的图标设计的集合，具有 2D 游戏艺术的风格，32K 超高清，详细的超现实主义，靛蓝和青铜，石头，精心制作的航天器，唐·奎克 --v 5

关键词组合

game icons assets sheet 5×5 of magic legendary, epic, games icons, rpg skills icons, honor of kings items icons, league of legends items icons, ability icon, flat pile gray background, octane rendering, 4k, raytracing --q 2 --niji 5

游戏图标资产表 5×5 魔法传奇，史诗，游戏图标，角色扮演游戏技能图标，《王者荣耀》物品图标，《英雄联盟》物品图标，能力图标，平堆灰色背景，Octane 渲染，4K，光线追踪 --q 2 --niji 5

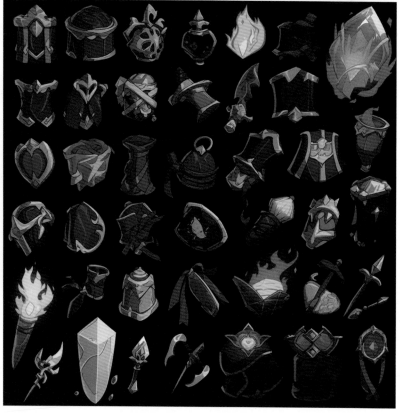

关键词组合

treasure chest crate for games on desktop or mobile, in the style of brian despain, dark white and gold, skull motifs, varied brushwork techniques, emerald --v 5

适用于桌面或移动游戏的宝箱，布瑞恩·德斯潘风格，深白色和金色，骷髅图案，各种笔触技巧，祖母绿 --v 5

关键词组合

an image of an emblem that shows a golden cross and blue star, packed with hidden details, loish, eye-catching tags, accurate and detailed --v 5

显示金色十字和蓝色星星的徽章图像，包含隐藏的细节，洛伊斯，醒目的标签，准确且详细 --v 5

关键词组合

lion shield of champions, in the style of 2d game art, dark gold and azure, streamlined forms, gongbi, eye-catching tags, meticulous linework precision, liquid metal --v 5

冠军狮子盾，2D 游戏艺术风格，暗金色和天蓝色，流线型的形式，工笔画，醒目的标签，细致的线稿精度，液态金属 --v 5

关键词组合

the league of legends logo with a dragon, in the style of luxurious geometry, ryan hewett, symmetrical harmony, 2d game art, layered texture, ue5, blink-and-you-miss-it detail --v 5

《英雄联盟》的龙形标志，豪华的几何风格，瑞安·赫维特，对称的和谐，2D 游戏艺术，分层纹理，UE5，眨眼就错过的细节 -- v 5

3.2 品牌类

关键词组合

the logo for game, in the style of cartoon realism, expressive character design, biopunk, explosive wildlife, trompe l'oeil technique, louis, dignified poses --v 5

游戏标志，卡通现实主义风格，富有表现力的角色设计，生物朋克，爆炸性野生动物，错视画技术，路易斯，庄严的姿势 --v 5

关键词组合

art bear logo, holding brushes in hands, anthropomorphic facial expression, wearing a small hat on top of the head, light maroon and light amber colors, cute cartoon logo, simple style, cartoon logo, flat color blocks, concise, flat logo, ip

艺术化的熊标志，手里拿着画笔，拟人化的面部表情，头顶戴了一顶小帽子，浅栗色和浅琥珀色，可爱的卡通标志，简约风格，卡通标志，平面色块，简洁，扁平标志，ip

关键词组合

a foxes logo for sport, in the style of light black and dark azure, playful cartoons, UE5, highly realistic, sharp & vivid colors, dark themes, luminous shadowing --v 5

狐狸运动标志，浅黑色和深蓝色风格，俏皮卡通，UE5，高度逼真，鲜明生动的色彩，黑暗主题，发光阴影 --v 5

关键词组合

lion logo, frontal head, open mouth, fierce face, dark orange and dark bronze colors, e-sports, badge, minimalism, cartoon style, trendy illustration, graphic logo

狮子标志，正面头部，张着嘴巴，凶狠的面貌，深橙色和深古铜色，电子竞技，徽章，极简主义，卡通风格，时尚插画，图形标志

关键词组合

duck logo, chubby and cute duck, chef hat on head, simple lines, simple color blocks, light dark red and white, anime style character design, kawaii manga style, q version cute style, disney style, comic illustration, cartoon signs, flat signs

鸭子标志，胖胖可爱的鸭子，头顶厨师帽，简单线条，简单色块，浅深红色和白色，动漫风格的角色设计，可爱的漫画风格，Q版可爱风，迪士尼风格，漫画式插画，卡通标志，扁平标志

关键词组合

animal logo, anthropomorphic cute panda holding coffee cup in hands, simple line, monochrome, white background, abstract style, cartoon logo, flat illustration, line art, minimalism

动物标志，拟人化的可爱熊猫，双手捧着咖啡杯，简单线条，单色，白色背景，抽象风格，卡通标志，平面插画，线条艺术，极简主义

关键词组合

vector and cat logo vector illustration, in the style of willem van aelst, precise nautical detail, octane render, celtic art, etruscan art, organic texture --v 5

矢量和猫标志矢量插图，威廉·凡·阿尔斯特风格，精确的航海细节，Octane 渲染，凯尔特艺术，伊特鲁里亚艺术，有机纹理 --v 5

关键词组合

a logo with a beautiful dog, vintage poster style, dark emerald and beige, painterly illustration, tonalist color scheme, art nouveau style --v 5

一个有漂亮的狗的标志，复古海报风格，深翡翠色和米色，强调色彩渲染的插图，有调性的配色方案，新艺术装饰风格 --v 5

关键词组合

botanical garden logo, in the style of nostalgic yearning, wild and daring, realist landscapes, light brown and green, outlined art --v 5

植物园标志，怀旧向往风格，狂野大胆，现实主义风景，浅棕绿色，勾勒艺术 --v 5

关键词组合

mountain view logo, highly detailed foliage, retro graphic design, simplified colors, otherworldly realism, tranquil garden landscape, strong linear elements --v 5

山景标志，高度细致的树叶，复古的图形设计，简化的颜色，超自然的现实主义，宁静的花园景观，强烈的线性元素 --v 5

关键词组合

logo illustration, delicate landscape, grey, dark details, highly detailed realism, retro poster design style, art nouveau --v 5

标志插图，精致的风景，灰色，黑暗的细节，高度细致的现实主义，复古海报设计风格，新艺术风格 --v 5

关键词组合

mountain view logo, woodblock print, high key style, graphic illustration, retro poster design, dark gold and beige, classic tattoo pattern, light brown and gold --v 5

山景标志，木版画，高色调风格，图形插图，复古海报设计，深金色和米色，经典文身图案，浅棕色和金色 --v 5

关键词组合

retro old farmer on round background with basket with fruits in hands, asian man, animated gif, punk, comic face, catchy label --v 5

复古老农在圆形背景上，手中的篮子里装着水果，亚洲人，GIF 动画，朋克，漫画面孔，引人注目的标签 --v 5

关键词组合

a logo for charlies hair salon, in the style of intense chiaroscuro portraits, dieselpunk, detailed engraving, 2d game art, bronze and amber, masculine, chalk --v 5

查理发廊的标志，采用强烈的明暗对比肖像风格，柴油朋克，详细的雕刻，2D 游戏艺术，青铜和琥珀，男性化，粉笔 --v 5

关键词组合

poster for captains hats badge, black ink on brown illustration, in the style of harmonious color fields, strong facial expression, navy, golden age aesthetics, maritime scenes, stock photo, raw character --v 5

船长帽徽海报，棕色插图上的黑色墨水，和谐色域风格，强烈的面部表情，海军，黄金时代美学，海上场景，库存照片，原始角色 --v 5

关键词组合

people logo, middle aged male with chef hat on head, front view, monochrome, block face, dark background, minimalism, cartoon logo, comic book style, flat illustration, line art --v 5

人物标志，中年男性头上戴着厨师帽，正面，单色，块面，深色背景，极简主义，卡通标志，漫画风格，平面插画，线条艺术 --v 5

关键词组合

cute, boy, blue suit, green hat, whole body, pop mart, leaf background, hd 8k --ar 3:4 --v 5

可爱的，男孩，蓝色的衣服，绿色的帽子，整个身体，泡泡玛特，叶子背景，高清 8K--ar 3：4 --v 5

关键词组合

illustration art illustration design by Shanghai creative studio , in the style of organic sculpting, light emerald and pink, miki asai, detailed character design, dutch baroque --ar 3:5 --q 2 --s 750 --v 5

插画艺术，插画设计，上海创意工作室设计，有机雕刻风格，浅翠绿色和粉色，浅井美纪，详细的人物设计，荷兰巴洛克风格 --ar 3：5 --q 2 --s 750 --v 5

关键词组合

an animated character with a big fruit, in the style of soft and dreamy tones, dolly kei, digital airbrushing, contest winner, childlike figures, low resolution --ar 3:4 --v 5

一个有大水果的动画角色，风格柔和梦幻，dolly kei 风格，数字喷枪，竞赛获胜者，童趣人物，低分辨率 --ar 3：4 --v 5

关键词组合

the little girl is wearing a goldfish hat, sitting on a doughnut, fishing, fantasy, small and cute, pop mart, next to the bubble emoticon, soft light, soft color, 3d icon clay rendering, blender 3d, soft background --ar 3:4 --v 5

小女孩戴着金鱼帽，坐在甜甜圈上，钓鱼，幻想，小而可爱，泡泡玛特，旁边的气泡表情符号，柔和的光线，柔和的颜色，3D 图标黏土渲染，Blender 3D，柔和的背景 --ar 3：4 --v 5

关键词组合

pop mart super cute girl ip, front view, full body, soft pastel color, baggy pants, headphones, clean pastel bright background, raking lightm, 3d, ultra detailed, c4d --ar 3:4 --v 5

泡泡玛特超可爱女孩 IP，正面图，全身，柔和的色彩，宽松的裤子，耳机，干净柔和的明亮背景，倾斜光线，3D，超详细，C4D --ar 3：4 --v 5

关键词组合

cartoon doll by jm, in the style of luminous and dreamlike scenes, cute cartoonish designs --ar 3:4 --v 5

JM 的卡通娃娃，风格是发光和梦幻般的场景，可爱的卡通设计 --ar 3：4 --v 5

关键词组合

the 3d illustration of an adorable little girl,melting, craft, James Jean, uhd image, bulbous --ar 3:4 --v 5

一个可爱的小女孩的 3D 插图，融化，工艺，詹姆斯·吉恩，超高清图像，球状的 --ar 3：4 --v 5

关键词组合

a cute cartoon girl wearing a strawberry helmet, in the style of delicate sculptures, dark white and light green, li tiefu, 32k uhd, loish, unique yokai illustrations --ar 3:4 --v 5

戴着草莓头盔的可爱卡通少女，精致的雕塑风格，深白色和浅绿色，李铁夫，32K 超高清，洛伊斯，独特的妖怪插画 --ar 3：4 --v 5

关键词组合

1923 exhibition bauhaus retro, nordic bauhaus retro, modern abstract geometric prints --ar 3:4 --v 5

1923 年包豪斯复古展，北欧包豪斯复古，现代抽象几何印刷品 --ar 3：4 --v 5

关键词组合

a painting showing two women with abstract shapes, in the style of ed mell, cubist portraiture, art deco, split toning, serene faces, symbolic composition, symbolic images --ar 1:1 --q 2 --s 750 --v 5

一幅展示两个抽象形状女人的画，埃德·梅尔的风格，立体派肖像画，装饰艺术，分裂色调，宁静的面孔，象征性构图，象征性图像 --ar 1：1 --q 2 --s 750 --v 5

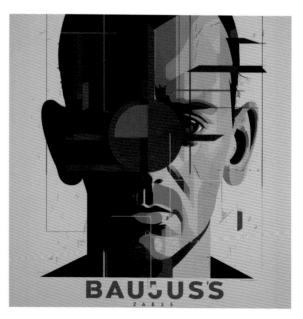

关键词组合

bauhaus movie poster --v 4 --v 5

包豪斯电影海报 --v 4 --v 5

关键词组合

bauhaus style poster for a philosophy convention --v 5

哲学大会的包豪斯风格海报 --v 5

关键词组合

bauhaus music logo --v 5

包豪斯音乐标志 --v 5

关键词组合

graphic design, minimalist music poster, design archive, typography poster, polish film poster, layout study, collectors edition --ar 3:4 --v 5

平面设计，极简主义音乐海报，设计档案，排版海报，波兰电影海报，布局研究，收藏版 --ar 3：4 --v 5

关键词组合

graphic design exhibition, music poster, design archive, typography poster, polish film poster, collectors edition --ar 3:4 --v 5

平面设计展，音乐海报，设计档案，排版海报，波兰电影海报，收藏版 --ar 3：4 --v 5

关键词组合

pop art: this style is characterized by the use of bold graphic elements and the incorporation of elements from popular culture, such as advertisements and comic books. an ai image generator could use this style to create colorful graphic compositions with a playful pop culture vibe. --ar 3:4 --v 5

波普艺术：这种风格的特点是使用大胆的图形元素，并结合流行文化的元素，例如广告和漫画书。AI 图像生成器可以使用这种风格来创建具有俏皮流行文化氛围的彩色图形作品。--ar 3：4 --v 5

关键词组合

pop art hotel sales pitch --v 5

波普艺术酒店的推销宣传 --v 5

关键词组合

marilyn monrose in the style of deborah azzopardi, dutch golden age, sky-blue and white, comic strip art, relief, 20th century art, kathrin longhurst --ar 1:1 --q 2 --s 750 --v 5

玛丽莲·梦露，黛博拉·阿佐帕迪的风格，荷兰黄金时代，天蓝色和白色，漫画艺术，浮雕，20世纪艺术，凯瑟琳·朗赫斯特 --ar 1：1 --q 2 --s 750 --v 5

关键词组合

a painting with an indian face, and many
other items, in the style of dark sky-blue
and dark yellow, mr. doodle, rene burri,
solarpunk, harmonious balance, ndebele art,
dariusz zawadzki --v 5

一幅印有印第安人面孔的画，还有许多其
他物品，风格为深天蓝色和深黄色，先生。
涂鸦，rene burri，太阳朋克，和谐平衡，
ndebele 艺术，达留什·扎瓦兹基 --v 5

关键词组合

artist, in the style of playful and colorful
depictions, gaston lachaise, jonathan
lasker, kadir nelson, multi-layered collages,
sculptural volumes, text-based art --v 5

艺术家，俏皮和丰富多彩的描绘风格，加
斯东·拉歇兹，乔纳森·拉斯克，卡迪尔·尼
尔森，多层拼贴画，雕塑体，基于文本的
艺术 --v 5

关键词组合

on my brows, black and blue abstract painting, a man carrying things for his family, in the style of surrealistic cartoons, dark yellow and orange, solarpunk, city portraits, light red and cyan, shiny eyes, stenciled iconography --v 5

在我的眉毛上，黑色和蓝色抽象画，一个男人为他的家人提着东西，超现实主义卡通风格，深黄色和橙色，太阳朋克，城市画像，浅红色和青色，闪亮的眼睛，模板图像 --v 5

关键词组合

piece with a geometric head and various colored shapes on top, in the style of paper sculptures, feminine pop art, object portraiture specialist, highly detailed figures, pont-aven school --v 5

带有几何头部并且顶部有各种彩色形状的作品，纸雕风格，女性波普艺术，对象肖像画专家，非常详细的数字，阿凡桥派 --v 5

关键词组合

the poster is light, in the style of dynamic and exaggerated facial expressions, vintage poster style, hyper-realistic representation, neo-dadaist, impasto frenzy, energetic gestures, tachist --v 5

海报很轻，采用动态和夸张的面部表情风格，复古海报风格，超现实表现，新达达主义，厚涂狂热，充满活力的手势，塔希派画家 --v 5

关键词组合

fancy print, in the style of poster art, gadgetpunk, vibrant exaggeration, arbeitsrat für kunst, patrick mchale, flat form, aggressive digital illustration --v 5

花式印刷，海报艺术风格，小工具朋克，充满活力的夸张，艺术工作委员会，帕特里克·麦克海尔，平面形式，激进的数字插图 --v 5

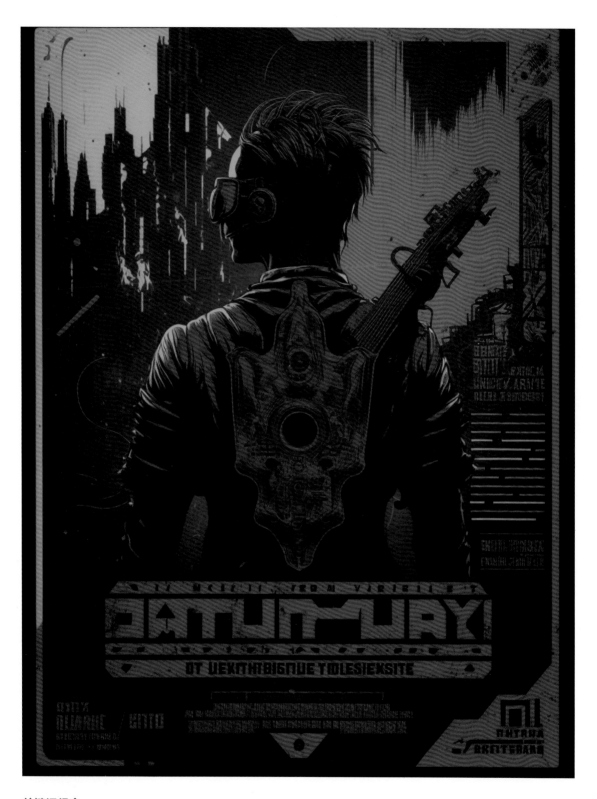

关键词组合

concert poster from a cyberpunk future society --v 4 --q 2 --ar 3:4 --v 5

来自赛博朋克未来社会的音乐会海报 --v 4 --q 2 --ar 3：4 --v 5

关键词组合

cyberpunk movie poster --ar 3:4 --v 4 --v 5

赛博朋克电影海报 --ar 3：4 --v 4 --v 5

关键词组合

by william morris --v 4 --v 5

威廉·莫里斯 --v 4 --v 5

关键词组合

william morris wallpaper, focused, sharp, crisp edges and lines, highly detailed with flowers and bird, high resolution, elegant, beautiful, colorful --v 5

威廉·莫里斯壁纸，目标明确，锐利，清晰的边缘和线条，细致程度高的花和鸟的细节，高分辨率，优雅，美丽，多彩 --v 5

关键词组合

design a william morris style wallpaper with plants native to texas --v 5

用得克萨斯州的原生植物设计威廉·莫里斯风格的墙纸 --v 5

关键词组合

detailed, high resolution, focused, clean, crisp sharp edges, beautiful, elegant, vibrant colors --v 5

详细，高分辨率，目标明确，干净，锐利的边缘，美丽，优雅，鲜艳的色彩 --v 5

关键词组合

repeating seamless art nouveau dragonfly pattern --v 4 --q 2 --v 5

重复的无缝新艺术蜻蜓图案 --v 4 --q 2 --v 5

关键词组合

repeating wallpaper pattern, in style of william morris, arts and crafts, stylized florals, vivid colors, intricate, elegant, cinematic, ornate + 4k + uhd + 3d + octane render, --v 4 --v 5

重复的壁纸图案，威廉·莫里斯的风格，艺术和手工艺，风格化的花朵，鲜艳的色彩，复杂，优雅，电影，华丽的 +4K+ 超高清 +3D+Octane 渲染，--v 4 --v 5

关键词组合

seamless mediterranean pattern, william morris, beautiful, intricate, ultra hd --v 5

无缝地中海图案，威廉·莫里斯，美丽，复杂，超高清 --v 5

关键词组合

william morris pattern design --v 5

威廉·莫里斯的图案设计 --v 5

3.3 摄影类

关键词组合

ox in the forest wallpaper, in the style of realistic portrait, realistic hyper-detailed rendering, hyper-realistic details, dark orange and red, softbox lighting, editorial illustrations --ar 1:1 --v 5

森林中的牛壁纸，逼真的肖像风格，逼真的超详细渲染，超现实的细节，深橙色和红色，柔光箱照明，编辑插图 --ar 1：1 --v 5

关键词组合

a horse in wild floral crown is in black and white, in the style of dark sky-blue and light orange, photo-realistic techniques, dark aquamarine and pink, playful innocence --ar 1:1 --v 5

戴着野花冠的马，黑白相间，深天蓝色和浅橙色风格，写实手法，深海蓝和粉色，俏皮童真 --ar 1：1 --v 5

关键词组合

a sheep dressed in flowers wears a crown, in the style of portraits with soft lighting,iso 200, matte photo, contest winner --ar 1:1 --v 5

披着花冠的绵羊，人像风格，柔光，ISO 200，磨砂照片，比赛获胜者 --ar 1：1 --v 5

关键词组合

the leopard stands next to some flowers, in the style of 8k, glowing colors, 4k, dark cyan, airbrush art, nature scenes, strong facial expression --ar 51:91 --v 5

豹子站在一些花旁边，8K风格，发光的颜色，4K，深青色，喷绘艺术，自然场景，强烈的面部表情 --ar 51：91 --v 5

关键词组合

majestic wolf, realistic photography, colorful background, detailed portrait, intricate details, rich colors, realistic style, front look, --ar 3:4 --v 5

雄伟的狼，逼真的摄影，彩色背景，详细的画像，错综复杂的细节，丰富的色彩，写实的风格，正面看，--ar 3：4 --v 5

关键词组合

panda from sichuan, in the style of andreas rocha, mikko lagerstedt, strong facial expression, 8k resolution, richard serra, realistic portrayal of light and shadow, detailed flora and fauna --ar 3:4 --v 5

来自四川的熊猫，安德烈亚斯·罗沙风格，米科·拉格斯泰特，强烈的面部表情，8K 分辨率，理查德·塞拉，光影的真实写照，详细的动植物 --ar 3：4 --v 5

关键词组合

3d rabbit lying on a dark background, in the style of realistic and hyper-detailed renderings, light amber, national geographic photo, high quality photo --v 5

3D 兔子躺在黑暗的背景上，以逼真的超详细渲染风格，浅琥珀色，国家地理照片，高质量照片 --v 5

关键词组合

an alpaca is in front of mountains and a llama, in the style of photorealistic portraits,mike campau, cartoony characters, backlight, hyperrealistic wildlife portraits, made of cheese --v 5

羊驼在山前，一只美洲驼，采用写实肖像风格，迈克·坎波，卡通人物，逆光，超写实野生动物肖像，由奶酪制成 --v 5

关键词组合

dog, blue scarf, in the style of illusory wallpaper portraits, 8k 3d, dark amber, hip hop aesthetics, color splash, american, painting --ar 1:1 --v 5

狗，蓝围巾，虚幻壁纸肖像风格，8K 3D，深琥珀色，嘻哈美学，色彩飞溅，美国的，绘画 --ar 1：1 --v 5

关键词组合

tabby cat in colorful sweater, in the style of realistic fantasy artwork, vray, photo-realistic techniques, charming realism --ar 1:1 --v 5

穿着五颜六色毛衣的虎斑猫，以逼真的奇幻艺术风格，VRay，逼真的技术，迷人的现实主义 --ar 1：1 --v 5

关键词组合

the stork bird wears flowers on his head, in the style of hyperrealistic landscapes, dark cyan and dark black, 8k resolution, vibrant palette, jungle punk, dreamlike creatures --ar 3:4 --v 5

鹳鸟头上戴着花，超现实主义风景画风格，深青色和深黑色，8K 分辨率，充满活力的调色板，丛林朋克，梦幻般的生物 --ar 3：4 --v 5

关键词组合

white swan swimming around in dark water, in the style of hyperrealistic wildlife portraits, hdr, dark white and light orange, wildlife photography, grandeur of scale, fujifilm velvia, brooding mood --ar 1:1 --v 5

白天鹅在黑暗的水中游来游去，超现实野生动物肖像风格，高动态范围成像，深白色和浅橙色，野生动物摄影，宏伟的规模，富士 Velvia，忧郁的心情 --ar 1 : 1 --v 5

关键词组合

image is a duck in pink flowers, in the style of realistic hyper-detailed rendering, uhd image, dark silver and light brown, free brushwork, texture exploration, norwegian nature --ar 1:1 --v 5

图像是粉红色花朵中的鸭子，采用逼真的超详细渲染风格，超高清图像，深银色和浅棕色，自由笔触，纹理探索，挪威自然 --ar 1 : 1 --v 5

关键词组合

¾,realistic photography, colorful background, detailed portrait, intricate details, rich colors, realistic style, front look, --v 5

¾,写实摄影，多彩背景，详细的画像，错综复杂的细节，丰富的色彩，写实风格，正面，--v 5

关键词组合

colorful parrot with colorful feathers standing on his own back, in the style of zbrush, textural explorations, pigeoncore, dark cyan and bronze, raw character, realistic, hyper-detail --ar 1:1 --v 5

背上带有彩色羽毛的鹦鹉，ZBrush 风格，纹理探索，鸽核，深青色和青铜色，原始角色，逼真，细致程度高 --ar 1：1 --v 5

关键词组合

3d bees on flower, bee in flower, bee with sting on sunflower, in the style of alchemical symbolism, orange and amber, emotive body language, documentary travel photography, intense and dramatic lighting, light orange and light gold, intel core --ar 3:4 --v 5

3D 蜜蜂在花上，花中的蜜蜂，向日葵上有刺的蜜蜂，以炼金术象征主义的风格，橙色和琥珀色，情绪化的肢体语言，纪实性的旅行摄影，强烈而戏剧性的灯光，浅橙色和浅金色，英特尔核心 --ar 3：4 --v 5

关键词组合

a bug is sitting on a leaf with large wings, in the style of hyper-realistic portraiture, dark sky-blue and light amber, large format lens, zbrush, bold and vibrant primary colors, hyper-realistic portraits, varied brushwork techniques --v 5

一只长着大翅膀的虫子坐在叶子上，超写实肖像风格，深天蓝色和浅琥珀色，大画幅镜头，ZBrush，大胆而充满活力的原色，超写实肖像，多种绘画技巧 --v 5

关键词组合

a butterfly is sitting on an orange flower with a dark background, in the style of vray tracing, photo-realistic techniques, renaissance-inspired chiaroscuro, intense and dramatic lighting, colorful storytelling, realistic, rubens --v 5

一只蝴蝶坐在一朵深色背景的橙色花朵上，采用 VRay 追踪风格，照片般逼真的技术，受文艺复兴时期启发的明暗对比，强烈而戏剧性的灯光，丰富多彩的故事，逼真，鲁本斯 --v 5

关键词组合

a grasshopper with wings standing on a green rock, in the style of vray tracing, realistic portrait drawings, photo-realistic techniques, mysterious jungle, rendered in cinema4d --v 5

一只长着翅膀的蚱蜢站在绿色的岩石上，采用 VRay 追踪风格，逼真的肖像画，照片般逼真的技术，神秘的丛林，在 Cinema 4D 中渲染 --v 5

关键词组合

dragonfly with small stout antenna, in the style of daz3d, hyperrealistic animal portraits, filip hodas, canon f-1, dark cyan and light amber, photo-realistic techniques, worthington whittredge --v 5

带有粗壮小触角的蜻蜓，Daz3D 风格，超写实动物画像，菲利普·霍达斯，佳能 F-1，深青色和浅琥珀色，逼真技术，沃辛顿·惠特里奇 --v 5

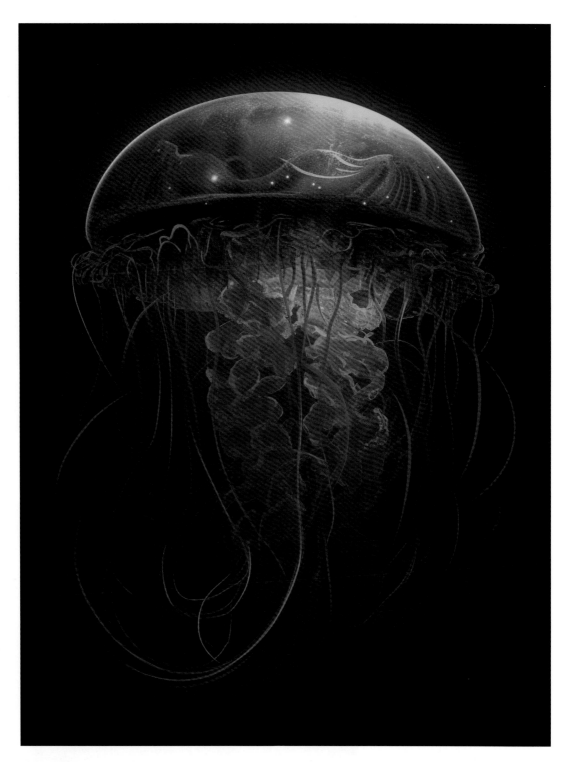

关键词组合

an underwater artwork of a jellyfish, in the style of luminescent color scheme, zbrush, undefined anatomy, realistic color palette, dark cyan and orange, soft lines, hurufiyya --ar 3:4 --v 5

水母的水下艺术品，发光配色方案风格，ZBrush，未定义的解剖，逼真的调色板，深青色和橙色，柔和的线条，胡鲁菲亚 --ar 3：4 --v 5

关键词组合

colorful fish in an aquarium wall art, in the style of uhd image, dark cyan and bronze, stockphoto, strong facial expression, dappled --ar 1:1 --v 5

水族馆墙壁艺术中的五颜六色的鱼，超高清图像风格，深青色和青铜色，库存照片，强烈的面部表情，斑驳的 --ar 1：1 --v 5

关键词组合

dolphin realistic rendering style, digital illustration, 8k resolution, soft shadows, ultra-detailed illustrations, strong chiaroscuro --v 5

海豚写实渲染风格，数字插画，8K 分辨率，柔和的阴影，超细致的插图，强烈的明暗对比 --v 5

关键词组合

a red and blue crab in shallow water, in the style of hyper-realistic sci-fi, colorful biomorphic forms, photobashing, dark turquoise and dark gold, indonesian art, national geographic photo, stock photo --ar 91:51 --v 5

浅水中的红蓝螃蟹，超写实科幻风格，丰富多彩的生物形态，照片拼接，深绿松石色和深金色，印度尼西亚艺术，国家地理照片，库存照片 --ar 91：51 --v 5

关键词组合

turtle sitting on the water, in the style of caras ionut, zbrush, detailed botanical illustrations, victor nizovtsev, close-up intensity, motion blur, ue5 --v 5

乌龟坐在水面上，卡拉斯·奥努特的风格，ZBrush，详细的植物插图，维克托·尼佐夫采夫，特写强度，运动模糊，UE5 --v 5

关键词组合

a woman in a dress and green hat, in the style of dreamlike portraiture, photobashing, uhd image, contrasting light and dark tones, floral, intense gaze, nature-inspired motifs --ar 3:4 --v 5

穿着裙子并且戴着绿色帽子的女人，采用梦幻般的肖像画风格，照片拼接，超高清图像，明暗色调的对比，花，强烈的凝视，受自然启发的图案 --ar 3：4 --v 5

关键词组合

an old man in costume standing with a lamp, in the style of dieselpunk, night photography, vivid portraiture, greg rucka, dramatic atmospheric perspective, romantic academia, classic academia --ar 3:4 --v 5

一个穿着戏服的老人拿着一盏灯站着，柴油朋克风格，夜间摄影，生动的画像，格雷格·鲁卡，戏剧性的大气视角，浪漫的学术界，经典的学术界 --ar 3：4 --v 5

关键词组合

male character in trench coat poses against light, in the style of cyberpunk imagery, nikon d850, 32k uhd, dragon art, emerald, larme kei, hyper-realistic oil --ar 3:4 --v 5

穿着风衣的男性角色在光线下摆姿势，以赛博朋克意象的风格，尼康 D850，32K 超高清，龙的艺术，祖母绿，Larme Kei 风格，超写实油画 --ar 3：4 --v 5

关键词组合

fresh chili pepper, seamless background, adorned with glistening droplets of water. top down view. shot using a hasselblad camera, iso 100. professional color grading. soft shadows. clean sharp focus. high-end retouching. food magazine photography. award winning photography. advertising photography. commercial photography. --ar 3:4 --v 5

新鲜的辣椒，无缝背景，装饰着闪闪发光的水滴。自上而下的视图。使用哈苏相机拍摄，ISO100。专业色彩分级。柔和的阴影。干净锐利的焦点。高端修饰。食品杂志摄影。获奖摄影作品。广告摄影。商业摄影。--ar 3：4 --v 5

关键词组合

fresh leek, seamless background, adorned with glistening droplets of water. top down view. shot using a hasselblad camera, iso 100. professional color grading. soft shadows. clean sharp focus. high-end retouching. food magazine photography. award winning photography. advertising photography. commercial photography. --ar 3:4 --v 5

新鲜的韭菜，无缝背景，点缀着闪闪发光的水滴。自上而下的视图。使用哈苏相机拍摄，ISO100。专业色彩分级。柔和的阴影。干净锐利的焦点。高端修饰。食品杂志摄影。获奖摄影作品。广告摄影。商业摄影。--ar 3：4 --v 5

关键词组合

fresh apple, seamless background, adorned with glistening droplets of water. top down view. shot using a hasselblad camera, iso 100. professional color grading. soft shadows. clean sharp focus. high-end retouching. food magazine photography. award winning photography. advertising photography. commercial photography. --ar 3:4 --v 5

新鲜的苹果，无缝背景，点缀着闪闪发光的水滴。自上而下的视图。使用哈苏相机拍摄，ISO100。专业色彩分级。柔和的阴影。干净锐利的焦点。高端修饰。食品杂志摄影。获奖摄影作品。广告摄影。商业摄影。--ar 3：4 --v 5

关键词组合

white peaches with drops of water on them, in the style of dark pink and dark orange, innovative page design, expressionist imagery, tabletop photography, zbrush --ar 1:1 --v 5

上面有水滴的白桃，深粉色和深橙色的风格，创新的页面设计，表现主义图像，桌面摄影，ZBrush --ar 1：1 --v 5

关键词组合

a group of orange sweets with water drops on dark paper, in the style of environmental portraiture, seapunk, chinapunk, shot on 70mm, dark silver and yellow, exotic --ar 1:1 --v 5

一组在深色纸上有水滴的橙色糖果，环境肖像画风格，海洋朋克，中国朋克，70mm 镜头，深银色和黄色，异国情调的 --ar 1：1 --v 5

关键词组合

a large group of green pears with the water droplet on a black background, in the style of textured compositions, dark gray and yellow, close-up shots, mundane materials, grit and grain, stockphoto, kitchen still life --ar 1:1 --v 5

黑色背景中一大堆绿色梨子带着水滴，纹理组合的构图风格，深灰色和黄色，特写镜头，平凡的材料，沙砾和谷物，库存照片，厨房静物 --ar 1：1 --v 5

关键词组合

fresh lime, seamless background, adorned with glistening droplets of water. top down view. shot using a hasselblad camera, iso 100. professional color grading. soft shadows. clean sharp focus. high-end retouching. food magazine photography. award winning photography. advertising photography. commercial photography. --v 5

新鲜酸橙，无缝背景，装饰着闪闪发光的水滴。从上往下看。使用哈苏相机拍摄，ISO 100。专业的色彩分级。柔和的阴影。干净锐利的焦点。高端修饰。食品杂志摄影。获奖的摄影作品。广告摄影。商业摄影。--v 5

关键词组合

fresh strawberrys, seamless background, adorned with glistening droplets of water. top down view. shot using a hasselblad camera, iso 100. professional color grading. soft shadows. clean sharp focus. high-end retouching. food magazine photography. award winning photography. advertising photography. commercial photography. --ar 3:4 --v 5

新鲜草莓，无缝背景，点缀着闪闪发光的水滴。自上而下的视图。使用哈苏相机拍摄，ISO100。专业色彩分级。柔和的阴影。干净锐利的焦点。高端修饰。食品杂志摄影。获奖摄影作品。广告摄影。商业摄影。--ar 3：4 --v 5

关键词组合

fresh watermelon, seamless background, adorned with glistening droplets of water. top down view. shot using a hasselblad camera, iso 100. professional color grading. soft shadows. clean sharp focus. high-end retouching. food magazine photography. award winning photography. advertising photography. commercial photography. --ar 3:4 --v 5

新鲜的西瓜，无缝背景，点缀着闪闪发光的水滴。自上而下的视图。使用哈苏相机拍摄，ISO100。专业色彩分级。柔和的阴影。干净锐利的焦点。高端修饰。食品杂志摄影。获奖摄影作品。广告摄影。商业摄影。--ar 3：4 --v 5

关键词组合

a single pink rose in a dark background, in the style of light orange and emerald, naturalist aesthetic, bella kotak, tranquil gardenscapes, samuel van hoogstraten, uhd image, nature-inspired --ar 3:4 --v 5

深色背景中的一朵粉红玫瑰，浅橙色和祖母绿色风格，自然主义美学，贝拉·科塔克，宁静的花园景观，超高清图像，自然灵感 --ar 3：4 --v 5

关键词组合

a purple flower with blurs around it, in the style of matthias haker, light gold and dark blue, sony fe 85mm f/1.4 gm, fanciful, dreamlike imagery, serene visuals, bold lines, vibrant color, dark white and blue --ar 1:1 --v 5

一朵周围模糊的紫色的花，马蒂亚斯·哈克尔的风格，浅金色和深蓝色，索尼FE 85mm f/1.4 GM，奇幻，梦幻般的意象，宁静的视觉效果，大胆的线条，鲜艳的色彩，深白色和蓝色 --ar 1 : 1 --v 5

关键词组合

a yellow rose sits in a rainy day location, in the style of light bronze and amber, photorealistic fantasies, tonalism genius, uhd image, nature-inspired art, nature-inspired, national geographic photo --ar 1:1 --v 5

雨天的一朵黄玫瑰，浅青铜色和琥珀色的风格，逼真的幻想，色调主义天才，超高清图像，受自然启发的艺术，受自然启发，国家地理照片 --ar 1 : 1 --v 5

关键词组合

flowers dripping in water on a plant, in the style of luminous spheres, organic nature-inspired forms, softbox lighting, light orange and light gold, humble charm, strong lighting contrasts, sharp focus --ar 1:1 --v 5

花朵在植物上滴水，以发光球体的风格，有机的，受自然启发的形式，柔光箱照明，浅橙色和浅金色，谦逊的魅力，强烈的灯光对比，锐利的焦点 --ar 1：1 --v 5

关键词组合

photograph orange flower on the dark background, in the style of nikon d850, delicate fantasy worlds, glistening, national geographic photo, tokina at-x 11-16mm f/2.8 pro dx ii, uhd image, nature-inspired art --ar 1:1 --v 5

拍摄深色背景上的橙色花，尼康 D850 风格，精致的幻想世界，闪闪发光，国家地理照片，图丽 AT-X 11-16mm f/2.8 PRO DX II，超高清图像，受自然启发的艺术 --ar 1：1 --v 5

3.4 建筑类

关键词组合

the building has large windows so people can to the interior, in the style of organic biomorphic forms, the stars art group , vray, hyper-realistic water, nature-inspired --ar 3:4 --q 2 --v 5

该建筑有大窗户，因此人们可以看到内部，采用有机生物形态，星星艺术团体，VRay，超逼真的水，受自然启发的风格 --ar 3：4 --q 2 --v 5

关键词组合

modern home living , in the style of unreal engine, densely textured or haptic surface, vray tracing, photo-realistic, geodesic structures, pseudo-realistic, vray --ar 16:9 --q 2 --v 5

现代家居生活，采用虚幻引擎风格，密集纹理或触觉表面，VRay 追踪，照片般的真实感、测地线结构，伪真实感，VRay --ar 16：9 --q 2 --v 5

关键词组合

palace in london city area, majestic, breathtaking, realistic --ar 16:9 --q 2 --v 5

伦敦市区的宫殿，雄伟、壮观、逼真 --ar 16：9 --q 2 --v 5

关键词组合

house in the woods, in the style of spatialism, minimalist sets, floating structures, romantic river scapes, clear edge definition, adventure themed, dark white and green --ar 3:4 --v 5

树林中的房子，空间主义风格，极简主义布景，漂浮的结构，浪漫的河景，清晰的边缘定义，冒险主题，深白色和绿色 --ar 3：4 --v 5

关键词组合

an abstract white building near a rocky lake, in the style of realistic and hyper-detailed renderings, forestpunk, vray, eco-friendly craftsmanship, sculpted, i can't believe how beautiful this is, timber frame construction --ar 3:4 --v 5

岩石湖边的一座抽象的白色建筑，现实和超详细的渲染风格，森林朋克，VRay，环保工艺，雕刻，我无法相信这是多么美丽，木结构建筑 --ar 3：4 --v 5

关键词组合

upscaling image #4 with a small house in front of a big lake with mountains in the background, in the style of futuristic elements, luxurious geometry, dark white and light white, asymmetrical balance, uhd image, bold structural designs --ar 3:4 --v 5

放大图片 #4，大湖前的小房子，背景是山，风格是未来主义元素，豪华的几何图形，深白色和浅白色，不对称的平衡，超高清图像，大胆的结构设计 --ar 3：4 --v 5

关键词组合

3d rendering of a futuristic building surrounded by several rocks, in the style of bauhaus-inspired designs, monochromatic white figures, haunting houses, asymmetrical framing, design by architects, minimalist designs, frequent use of diagonals --ar 3:4 --v 5

被几块岩石包围的未来主义建筑的 3D 效果图，包豪斯风格的设计，单色的白色图形，使人难忘的房子，不对称的框架，建筑师的设计，极简主义设计，经常使用对角线 --ar 3：4 --v 5

关键词组合

the inside of a large building in an apartment, in the style of naturalistic rendering, warm tonal range, light-filled, varying wood grains, danish golden age, luxurious, vibrant airy scenes --ar 3:4 --s 250 --v 5

公寓内的大型建筑内部，自然主义渲染风格，暖色调范围，光线充足，变化的木纹，丹麦黄金时代，奢华，充满活力的通风场景 --ar 3：4 --s 250 --v 5

关键词组合

a black book case and a white dresser, in the style of unreal engine, vintage poster design, dark gray, 4k, dry wit humor, light amber and white --s 250 --v 5

黑色书柜和白色梳妆台，虚幻引擎风格，复古海报设计，深灰色，4K，冷幽默，浅琥珀色和白色 --s 250 --v 5

关键词组合

a furniture unit featuring a tv and a small plant, in the style of varying wood grains, ominous vibe, orderly symmetry, use of screen tones, high detail, jazzy interiors --s 250 --v 5

一个家具单元，配有电视和小植物，风格多样的木纹，不祥的氛围，有序的对称，使用屏幕色调，详细的细节，爵士风格的内饰 --s 250 --v 5

关键词组合

a wardrobe with many towels hanging in it, in the style of realistic chiaroscuro lighting, daz3d, classical symmetry, timeless nostalgia, vignettes of paris, packed with hidden details, layered veneer panels --s 250 --v 5

一个挂着许多毛巾的衣柜，采用逼真的明暗对比灯光风格，Daz3D，经典对称，永恒的怀旧，巴黎的小插曲，充满隐藏的细节，分层贴面板 --s 250 --v 5

关键词组合

a yellow chair, lamp and a tall plant on a yellow background, in the style of daz3d, orderly arrangements, collecting and modes of display, dark orange, realistic and hyper-detailed renderings --s 250 --v 5

黄色椅子，黄色背景中的灯和高大的植物，Daz3D 风格，有序地排列，收集和展示方式，深橙色，逼真和超详细的渲染 --s 250 --v 5

关键词组合

materials, in the style of subtle pastel hues, rounded shapes, warm color palette, contemporary ceramics, silhouette lighting, pink and beige, ultra detailed --ar 3:4 --v 5

用白色材料制作的现代客厅，采用微妙的淡雅色调，圆润的形状，温暖的色调，当代陶瓷，剪影照明，粉色和米色，超详细的 --ar 3：4 --v 5

关键词组合

living room with light colored furniture, in the style of soft and rounded forms, porcelain, playful use of line, pink and beige, contoured shading, modular design --ar 3:4 --v 5

客厅配有浅色家具，风格柔和，圆润的形式，瓷器，灵活运用线条，粉色和米色，轮廓阴影，模块化设计 --ar 3：4 --v 5

关键词组合

a living room with windows overlooking the mountains, in the style of chinese tradition, vray tracing, 32k uhd, gongbi, sculptural landscapes, mountainous vistas --ar 3:4 --v 5

一间带窗户的客厅，可俯瞰群山，中国传统风格，VRay 追踪，32K 超高清，工笔，雕塑风景，山景 --ar 3：4 --v 5

关键词组合

modern lounge furniture and styling , in the style of realistic chiaroscuro lighting, light orange and gray, marble, high quality photo, oriental minimalism, photorealistic rendering, dark gray and gray --ar 3:4 --v 5

现代休闲家具和造型，采用逼真的明暗对比，浅橙色和灰色，大理石，高品质照片，东方极简主义，逼真的渲染，深灰色和灰色 --ar 3：4 --v 5

关键词组合

3d cartoon anime house by design, in the style of influenced by ancient chinese art, soft color palette, high-angle, meticulously detailed, pixelation --s 250 --v 5

3D 卡通动漫屋设计，风格受中国古代艺术影响，柔和的调色板，高角度，细致，像素化 --s 250 --v 5

关键词组合

hassakeh isometric building town 2d illustration, in the style of tibor nagy, asian-inspired, jakub schikaneder, dark cyan and beige, stark and unfiltered, strong use of color, thai art --s 250 --niji 5

哈萨亚等距建筑城镇 2D 插图，采用蒂博尔·纳吉风格，亚洲风格，雅各布·席卡内德，深青色和米色，鲜明且未经过滤，强烈的色彩，泰国艺术 --s 250 --niji 5

关键词组合

3d illustration of an old town and buildings, in the style of asian-inspired, modular constructivism, the stars art group, isometric, stark realism, clean and simple designs --s 250 --niji 5

老城区和建筑物的 3D 插图，采用亚洲风格，模块化建构主义，星星艺术团体，等距，赤裸裸的现实主义，干净简单的设计 --s 250 --niji 5

关键词组合

3d model of a chinese temple, in the style of 2d game art, soft, muted tones, detailed character illustrations, street scenes with vibrant colors, playful use of shapes, design/architecture study, sopheap pich --s 250 --v 5

中国寺庙的 3D 模型，2D 游戏艺术风格，柔软，柔和的色调，详细的人物插图，色彩鲜艳的街景，有趣的形状使用，设计 / 架构研究，索皮·比 --s 250 --v 5

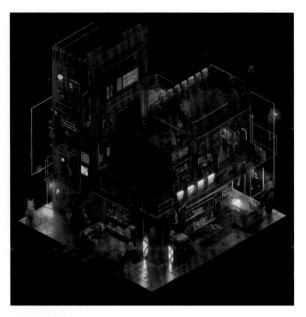

关键词组合

a classical two-story b&b on the lake, with lanterns, a ferry boat on the side, 2.5d isometric, cyberpunk city --v 5

湖边古典二层民宿，挂着灯笼，旁边有一艘摆渡船，2.5D 等距，赛博朋克城市 --v 5

关键词组合

two-level bar in the city, 2.5d isometric, cyberpunk city --v 5

城市中的两层酒吧，2.5D 等距，赛博朋克城市 --v 5

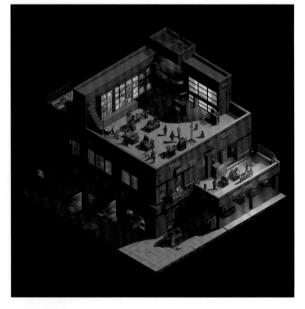

关键词组合

two-story fish-selling store in the city, 2.5d isometric, cyberpunk city --v 5

城市中的两层卖鱼店，2.5D 等距，赛博朋克城市 --v 5

关键词组合

two-story school in the city, 2.5d isometric, cyberpunk city --v 5

城市中的两层的学校，2.5D 等距，赛博朋克城市 --v 5

关键词组合

3d house with pool, 3d render illustration vector, in the style of muted, earthy tones, meticulous design, birds-eye view, daz3d, metropolis meets nature, design by architects, blocky --s 250 --v 5

带游泳池的 3D 房子，3D 渲染插图矢量，柔和的风格，朴实的色调，精心设计，鸟瞰图，Daz3D，大都会与自然相遇，建筑师设计，块状 --s 250 --v 5

关键词组合

3d model of a residential building on an island, in the style of cyril rolando, vacation dadcore, rich and tonal, enigmatic tropics, beige and aquamarine, arman manookian, bold yet graceful --s 250 --v 5

岛上住宅建筑的 3D 模型，西里尔·罗兰多风格，度假老爹风，丰富而有调性，神秘的热带，米色和海蓝宝石，阿尔曼·马努克，大胆而优雅 --s 250 --v 5

关键词组合

a cottage in the middle of an island in the ocean, in the style of rendered in cinema4d, art deco-inspired, tropical baroque, high-angle, muted tones, seapunk, intricate architectures --s 250 --v 5

一座位于海洋岛屿中央的小屋，采用 Cinema 4D 渲染风格，装饰艺术风格，热带巴洛克风格，高角度，柔和色调，海洋朋克，错综复杂的建筑 --s 250 --v 5

关键词组合

simple house with a swimming pool on a hill, in the style of light white and turquoise, rough clusters, hyper-realistic details, mediterranean-inspired, sculpted, streetscape, lively tableaus --s 250 --v 5

山上带游泳池的简单房子，浅白色和绿松石风格，粗糙聚集，超写实的细节，地中海风格，雕刻，街景，生动的画面 --s 250 --v 5

关键词组合

3d bedroom with cactus plants and chairs, in the style of 2d game art, darkly detailed, light navy and purple, studyplace, colorful palette, found objects, isometric --s 250 --niji 5

带仙人掌植物和椅子的 3D 卧室，采用 2D 游戏艺术风格，深色细节，浅海军蓝和紫色，书房，彩色调色板，拾来的材料，等距 --s 250 --niji 5

关键词组合

3d illustration of a orange interior in a tiny kitchen, in the style of playfully intricate, bold graphic illustrations, subtle lighting contrasts, isometric, dark gray and green, found-object-centric, soft gradients --s 250 --v 5

小厨房中橙色内饰的 3D 插图，风格复杂，大胆的图形插图，微妙的灯光对比，等距，深灰色和绿色，以发现的对象为中心，柔和的渐变 --s 250 --v 5

关键词组合

a bedroom with a desk and plants is pictured in a 3d illustration, in the style of dark cyan and violet, hyper-detailed illustrations, isometric, atmospheric color washes, i can't believe how beautiful this is, illustration, cute and colorful --s 250 --niji 5

在 3D 插图中描绘一间带书桌和植物的卧室，采用深青色和紫罗兰色的风格，超详细插图，等距，有水洗色氛围的，我不敢相信这是多么美丽，插图，可爱和多彩 --s 250 --niji 5

关键词组合

isometric tv room design, in the style of rhads, vintage aesthetics --s 250 --v 5

等距电视室设计，采用艺术家 Rhads 的风格，复古美学 --s 250 --v 5

3.5 模型类

关键词组合

a beautiful wine packaging with a label of dark dramatic sinister looking anime landscape, intricate detail, wallpaper, volumetric lighting, in the style of van gogh and beksinksi --ar 3:4 --v 5

一个漂亮的带有黑暗戏剧性险恶动画风景标签的葡萄酒包装，复杂的细节，墙纸，体积照明，凡·高和贝克辛斯基的风格 --ar 3：4 --v 5

关键词组合

3d design of the luxury packaging of a wine bottle with earth, vines and grapes with the colors dark blue--v 5

带有泥土、葡萄和葡萄的酒瓶的豪华包装的 3D 设计，颜色为深蓝色 --v 5

关键词组合

a red wine package in gold and black tones with golden text and icons, as well as grape-themed illustrations. overall, the packaging design reflects a feeling of elegance and luxury. ultra hd, super delicate pattern. --v 5

金色和黑色色调的红酒包装，带有金色文字和图标，以及以葡萄为主题的插图。总体而言，包装设计体现了优雅和奢华的感觉。超高清，超级精致的图案。--v 5

关键词组合

wine design briefs packaging and design, in the style of surreal 3d landscapes, light emerald and dark bronze, frostpunk, speedpainting, clear and precise bird art, realistic yet stylized, dau al set --v 5

葡萄酒设计简介包装和设计，在超现实的 3D 景观风格，浅绿宝石和深青铜，霜冻朋克，速度画，清晰和精确的鸟类艺术，现实但风格化，达乌塞 --v 5

关键词组合

wine lable design, highly detailed, hyperrealistic fashion shot, unreal engine, octane render, 8k --v 5

酒标设计，高度细致，超逼真的时尚镜头，虚幻引擎，Octane 渲染，8K --v 5

关键词组合

luxurious wine bottle designs --ar 3:4 --v 5

豪华酒瓶设计 --ar 3：4 --v 5

关键词组合

liquor bottle packaging design mood board, super detailed, halo infinite style --v 5

酒瓶包装设计情绪板，超详细，《光环：无限》风格 --v 5

关键词组合

create a fancy label for a tequila brand, mexican style --v 5

为龙舌兰酒品牌制作精美标签，墨西哥风格 --v 5

关键词组合

beer bottles design inspired in miguel angel --v 5

啤酒瓶设计的灵感来自于米格朗·格尔 --v 5

关键词组合

create a cover with itaipava beer with mostly yellow colors. --v 5

用以黄色为主的 Itaipava 啤酒制作封面。--v 5

关键词组合

a collection of unique craft brewery beer cans, front views --v 5

一系列独特的精酿啤酒罐，前视图 --v 5

关键词组合

alcohol can design, packaging design, highly graphical, bold colors --v 5

酒罐设计、包装设计、高度图形化、大胆的色彩 --v 5

关键词组合

belacan beaker packaging by chibi and braham aston, in the style of rebecca guay, bold traditional, germanic art, avian-themed, dark gold, unreal engine, becky cloonan --ar 1:1 --v 5

由 Chibi 和 Braham Aston 设计的峇拉煎烧杯包装，采用 Rebecca Guay 的风格，大胆的传统，日耳曼艺术，以鸟类为主题，暗金色，虚幻引擎，贝基·克洛南 --ar 1：1 --v 5

关键词组合

packaging design, popular can, natural feel, vintagepunk, samurai, blue and golden details, james jean inspiration, clear shape and design, 8k --v 5

包装设计，流行的罐子，自然的感觉，复古朋克，武士，蓝色和金色的细节，詹姆斯·简的灵感，清晰的形状和设计，8K --v 5

关键词组合

moodboard, premium luxury skincare sustainable packaging for gen z consumers, realistic --ar 3:4 --v 5

情绪板，高级奢华护肤品为 Z 世代消费者提供的可持续包装，现实的 --ar 3：4 --v 5

关键词组合

moodboard, premium luxury skincare sustainable packaging for gen z consumers, realistic --v 5

情绪板，高级奢华护肤品为 Z 世代消费者提供的可持续包装，现实的 --v 5

关键词组合

eco friendly, green, facial cosmetics brand --v 5

生态友好，绿色，面部彩妆品牌 --v 5

关键词组合

collection of skincare products including serums, toners and creams --v 5

护肤品系列，包括精华液、爽肤水和面霜 --v 5

关键词组合

servicio de suscripción de cajas de comida saludable y personalizadas --v 5

健康个性化食品盒订阅服务 --v 5

关键词组合

packaging design image --v 5

包装设计形象 --v 5

关键词组合

packaging design for healthy food trays --v 5

健康食品托盘的包装设计 -- v 5

关键词组合

dinner food creative packaging design --v 5

晚餐食品创意包装设计 --v 5

关键词组合

food pouch packaging design --v 5

食品袋包装设计 --v 5

关键词组合

food pouch packaging --v 5

食品袋包装 --v 5

关键词组合

foods packaging design,modern, exquisite, super detail, on behance --v 5

食品包装设计，现代，精致，超级细节，在 Behance 网站 --v 5

关键词组合

packaging en doypack, original et créatif, d'une compote de pomme --v 5

自立袋包装，原创且富有创意，苹果蜜饯 --v 5

关键词组合

supplement label design --v 5

补充标签设计 --v 5

关键词组合

red fat burner packaging, modern and sporty design --v 5

红色脂肪燃烧器包装，现代和运动型设计 --v 5

关键词组合

custom whey product --v 5

定制的乳清产品 --v 5

关键词组合

personalied vitamins for wellness --v 5

个性化健康维生素 --v 5

关键词组合

organic juice brand concept --v 5

有机果汁品牌理念 --v 5

关键词组合

juice beverage packaging design --v 5

果汁饮料包装设计 --v 5

关键词组合

beautiful botles for mango juices --v 5

漂亮的芒果汁瓶 --v 5

关键词组合

2235 Lamborghini, Ken Block gymkhana hillclimb prototype, particularly exaggerated shape, twin turbo, 4wd, hoonigan racing, big spoilers, low rakish stance, pikes peak unlimited, futuristic, science fiction, octane render --ar 16:9 --v 5 --q 2

2235 兰博基尼，肯·布洛克·金卡纳爬坡原型，特别夸张的形状，双涡轮增压，4WD，Hoonigan 赛车，大扰流板，低倾斜姿态，派克峰不受限制，未来主义，科幻，Octane 渲染 --ar 16：9 --v 5 --q 2

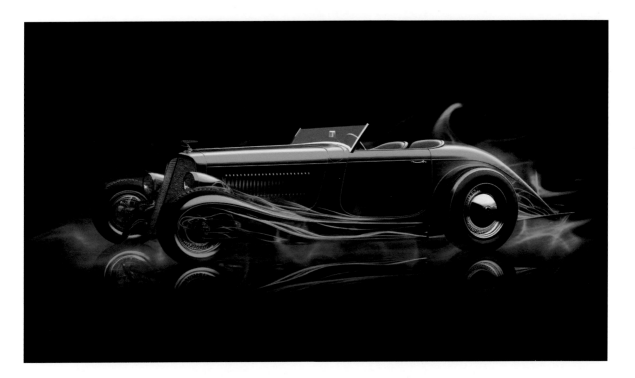

关键词组合

perfect equal mix of creative automotive brand design and luscious hot rod bombshell roadster brand design, digital art, clean, minimalist, highly detailed, futuristic, science fiction, hyperrealistic, editorial photography, hd, 3d, designer style, joy, beautiful, luminous, luminosity, tone mapping, front view, refreshing, studio lighting, side view, smoke, silver green, gold maroon, speed, tracing global illumination, ray tracing reflections, octane render --v 5 --ar 16:9

完美的混合创意汽车品牌设计和诱人的热棒炸弹跑车品牌设计, 数字艺术, 清洁, 简约, 非常详细, 未来主义, 科幻小说, 超现实主义, 编辑摄影, 高清, 3D, 设计师风格, 快乐, 美丽, 发光, 发光, 音调映射, 前视图, 清新, 工作室照明, 侧视图, 烟雾, 银绿色, 金栗色, 速度, 追踪全局照明, 光追踪反射, Octane 渲染 --v 5 --ar 16：9

关键词组合

the terminator design dyson supersonic hair dryer, technical detail, product photography, cinematic lighting, hyper detailed, photorealistic, hdr10, 8k, octane render --ar 3:4 --v 5

终结者设计的戴森超音速吹风机，技术细节，产品摄影，电影级灯光，超细节，逼真，高动态范围成像 10，8K，Octane 渲染 --ar 3：4 --v 5

关键词组合

a collection of spectacular kitchen appliances designed by colani, Sense of technology, futurism, transparent texture, 4k --v 5 --q 2 --ar 16:9

由科拉尼设计的一系列壮观的厨房用具，科技感，未来主义，透明质地，4K--v 5 --q 2 --ar 16：9

关键词组合

a collection of spectacular kitchen appliances designed by colani, 4k --v 5 --q 2 --ar 16:9

由科拉尼设计的壮观的厨房电器系列，4K --v 5 --q 2 --ar 16：9

关键词组合

a male razor, metal texture, and exquisite, jonathan paul ive style, 8k, octane render --ar 3:4 --v 5

男用剃须刀，金属质感，精致，乔纳森·伊夫风格，8K，Octane 渲染 --ar 3：4 --v 5

关键词组合

rice cooker by jonathan paul ive , fine details, minimalist style, white background, 8k, octane render -- v 5 --ar 3:4

乔纳森·伊夫设计的电饭锅，精致的细节，简约风格，白色背景，8K，Octane 渲染 --v 5 --ar 3：4

关键词组合

dyson air purifier cool tp01, science fiction, futurism, crystal texture,robotic, render on white background --v 5 --ar 3:4

戴森空气净化器 COOL TP01，科幻小说，未来主义，水晶纹理，机器人，白色背景上的效果图 --v 5 --ar 3：4

关键词组合

minimalistic dishwasher designed by apple, bauhaus, dieter rams, studio lighting, 8k, super high detailed, gorgeous 3d render --v 5 --ar 3:4

苹果公司设计的极简主义洗碗机，包豪斯，迪特尔·拉姆，工作室照明，8K，超高细节，华丽的 3D 渲染 --v 5--ar 3：4

关键词组合

product view, gamepad, delicate texture, sci-fi, futuristic, technological, product lighting, rich detail, design, industrial --ar 3:4 --q 2

产品视图，游戏手柄，精致的纹理，科幻，未来主义，技术，产品照明，丰富的细节，设计，工业 --ar 3：4 --q 2

关键词组合

mouse, neon colour, translucent melt, designed by dieter rams, high detail, 4k, industrial, bauhaus, design, white background, studio lighting，--s 750 --v 5 --q 2 --ar 16:9 --q 2

鼠标，霓虹灯色，半透明的熔体，由迪特·拉姆斯设计，高细节，4K，工业，包豪斯，设计，白色背景，工作室照明，--s 750 --v 5 --q 2 --ar 16：9 --q 2

关键词组合

close up shot, product-view, earphone, with a magical and luxurious quality, colorful light, magazine photography, advertorial, photorealistic,professional photo shootcinematic, photo realistic, ultra detailed, tone mapping, ray tracing, hd, 16k --ar 16:9 --v 5 --q 2

特写镜头，产品视图，耳机，具有神奇和豪华的质量，丰富多彩的光线，杂志摄影，广告，逼真，专业照片摄影，照片逼真，超细节，色调映射，光线追踪，高动态范围成像，16K --ar 16：9 --v 5 --q 2

关键词组合

close-up, product view, camera, gorgeous texture, tech feel, studio lighting, rich details, design, industrial --ar 16:9 --q 2

特写，产品视图，相机，华丽的纹理，科技感，演播室照明，丰富的细节，设计，工业 --ar 16：9 --q 2

关键词组合

in-studio shooting, product-view, macbook pro, with a magical and luxurious quality. colorful light, magazine photography, advertorial, photorealistic, uitra detailed,ray tracing, hdr, 16k --ar 16:9 --q 2

在工作室拍摄，产品展示，MacBook pro，具有神奇和豪华的质量。丰富多彩的光线，杂志摄影，广告，逼真，超详细，光线追踪，高动态范围成像，16K --ar 16：9 --q 2

关键词组合

designer chair, upholstery, design magazine, glass, gold --q 2 --v 5 --ar 3:4

设计师椅，软体家具，设计杂志，玻璃，黄金 --q 2 --v 5 --ar 3：4

关键词组合

futuristic furniture, wendell castle ::2:: ross lovegrove ::1:: mire lee 072 --v 5 --ar 3:4

未来主义家具，温戴尔·卡索 ::2:: 洛斯·拉古路夫 ::1:: 李美来 072 --v 5 --ar 3：4

关键词组合

high design of an organic wood desk using biomimicry of natural design --v 5 --ar 3:4

利用自然设计的生物仿生学对有机木桌进行高度设计 --v 5 --ar 3：4

关键词组合

modern dining table design with a fusion of glass and wood, a premium texture --ar 16:9

融合了玻璃和木材的现代餐桌设计，优质质感 --ar 16：9

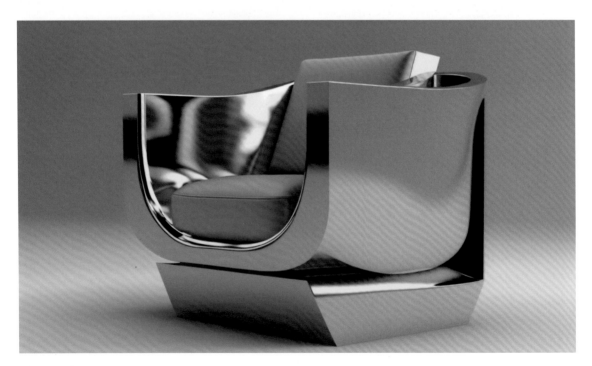

关键词组合

simple armchair design. precious metals, gold, award winner --ar 16:9 --stylize 500 --v 4 --q 2 --s 20000

简单的扶手椅设计。贵金属，黄金，获奖者 --ar 16：9 --stylize 500 --v 4 --q 2 --s 20000

3.6 界面类

关键词组合

a set of blue ui kits for apps, in the style of unified ui kit for responsive websites and apps, with different types of colors, in the style of dark white and light violet, gothic dark and moody tones, willem haenraets, matte photo, rounded shapes, multilayered, tibor nagy --ar 3:4

一套适用于应用程序的蓝色 UI 套件，采用适用于响应型网站和应用程序的统一 UI 套件的风格，具有不同类型的颜色，有深白色和浅紫色，哥特式的深色和忧郁色调，威廉·汉雷斯，亚光照片，圆形，多层，蒂博尔·纳吉 --ar 3：4

关键词组合

a gallery of 24 beautiful icons for an ios app, minimalist, simple, 2d --ar 1:1 --q 2

一个包含 24 个 iOS 应用程序精美图标的图库，简约，简单，2D --ar 1：1 --q 2

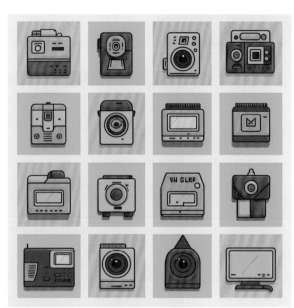

关键词组合

an assortment of icons with flat backgrounds, in the style of kodak vision3, dark yellow and aquamarine, cliff chiang, light red and cyan, vintage aesthetics, simplified line work, low bitrate --ar 1:1 --q 2

各种平面背景的图标，采用柯达 Vision3 的风格，深黄色和海蓝色，克里夫·江，浅红色和青色，复古美学，简化线条，低比特率 --ar 1:1 --q 2

关键词组合

logo, icon set, ui/ux, retro, colorful, modern, simplicity, transparent background --ar 1:1 --q 2

标志，图标集，UI/UX，复古，多彩，现代，简约，透明背景 --ar 1：1 --q 2

关键词组合

web and app icons line design --ar 1:1 --q 2

网络和应用程序图标线条设计 --ar 1：1--q 2

关键词组合

with a ui user interface with information about mountains and mountains, in the style of dark and brooding designer, realistic color palette, serge marshennikov, suburban ennui capturer, whimsical elements, large canvas sizes, realistic yet stylized --ar 1:1 --q 2

带有关于山脉和山脉信息的 UI 用户界面，采用黑暗和沉思的设计师风格，逼真的调色板，谢尔盖·马什尼科夫，郊区的倦怠捕捉器，异想天开的元素，大画布尺寸，逼真但程式化 --ar 1：1 --q 2

关键词组合

a user interface theme with a man and other items scattered around, in the style of striking contrasts of light and dark, realism with fantasy elements, aleksi briclot, unique character design, soft color fields, multiple screens, strong diagonals --ar 1:1 --q 2

一个用户界面主题，包含一个人和其他散落在周围的物品，采用鲜明的明暗对比风格，带有幻想元素的现实主义，阿莱克西·布里科洛，独特的角色设计，柔和的色域，多个屏幕，强烈的对角线 --ar 1：1 --q 2

关键词组合

an interactive dashboard with all the different displays, in the style of dark bronze and teal, vintage minimalism, realistic details, dark black and yellow, low bitrate, mechanical designs, use of precious materials --ar 1:1 --q 2

一个交互式仪表板，具有各不相同的显示屏，采用深青铜和蓝绿色风格，复古极简主义，逼真的细节，深黑色和黄色，低比特率，机械设计，使用珍贵材料 --ar 1：1--q 2

关键词组合

a dark web dashboard with an animal character, in the style of realistic usage of light and color, anna dittmann, memphis design, hyperrealistic details, vibrant and colorful, heavy texture, planar art --ar 1:1 --q 2

一个带有动物角色的暗网仪表板，采用光线和颜色的现实使用风格，安娜·迪特曼，孟菲斯设计，超逼真的细节，充满活力、丰富多彩，厚重的纹理，平面艺术 --ar 1：1--q 2

关键词组合

ui control panel with a number of elements showing data, in the style of light gray and red, minimalist black and white, light yellow and orange, juxtaposition of hard and soft lines, light white and pink, dark cyan and gray, flat shapes --ar 1:1 --q 2

具有多个显示数据元素的 UI 控制面板，风格为浅灰色和红色，极简主义的黑色和白色，浅黄色和橙色，硬线和软线并置，浅白色和粉色，深青色和灰色，扁平形状 --ar 1 : 1--q 2

关键词组合

a design of a modern app screen, made from abstract layered shapes, in the style of light green and cyan, industrial and product design, data visualization, rounded forms, light purple and light cyan, spot metering, sketch-like --ar 1:1 --q 2

一个现代应用程序屏幕的设计，由抽象的分层形状制成，呈浅绿色和青色风格，工业和产品设计，数据可视化，圆形，浅紫色和浅青色，点测光，草图状 --ar 1 : 1--q 2

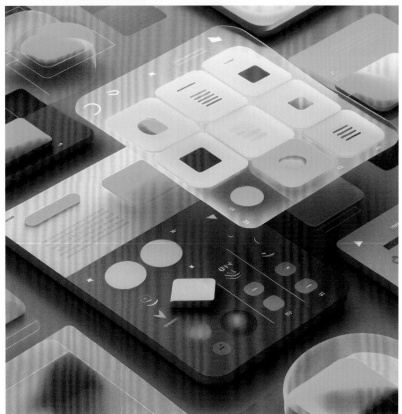

关键词组合

redshift ui kit, oscarhohenstein, in the style of soft pastel landscapes, transparent layers, pastel color scheme, realistic details, multi-layered color fields, mountainous vistas, soft shading --ar 1:1 --q 2

使用红移奥斯卡海亨斯坦 UI 套件，具有柔和蜡笔风格的风景，透明的层次，柔和的配色方案，逼真的细节，多层色场，多山的远景，柔和的明暗处理风格 --ar 1 : 1 --q 2

关键词组合

a photo of a website that has several screens and buttons, in the style of light beige and azure, colorful gradients, flat forms, dark azure and red, datamosh, fine detailed, modular design --ar 1:1 --q 2

有几个屏幕和按钮的网站照片，采用浅米色和天蓝色风格，色彩渐变，水平外表，深蓝色和红色，Datamosh 插件，精细细节，模块化设计 --ar 1 : 1 --q 2

关键词组合

an icon set that has different things in different colors, in the style of retro visuals, dark yellow and light cyan, kodak ektar, earthy textures, light red and dark gray, analog, realistic details --ar 1:1 --q 2

用不同颜色表示不同东西的一个图标集，复古视觉风格，深黄色和浅青色，柯达 EKTAR，泥土纹理，浅红色和深灰色，模拟，逼真的细节 --ar 1 : 1 --q 2

关键词组合

revolution game icon set, in the style of moody color schemes, thick texture, otherworldly fun, dark themes, soft pastels, light gold and dark cyan, leica r3 --ar 1:1 --q 2

"革命"游戏图标集，采用忧郁的配色方案，厚重的纹理，超凡脱俗的乐趣，深色主题，柔和的粉彩，浅金色和深青色，徕卡 R3 --ar 1 : 1 --q 2

关键词组合

icon design, light texture, glow, dribble, 3d, frosted glass effect, 3d, ui, ux, --upbeta --ar 1:1 --q 2

图标设计，光线纹理，发光，漂移，3D，磨砂玻璃效果，3D，UI，UX，--upbeta--ar 1：1 --q 2

关键词组合

a collection of colored and stylized icons, in the style of realistic landscapes with soft, tonal colors, light pink and light amber, social media art, m42 mount, rolleiflex original, vladimir kush, light white and light gray --ar 1:1 --q 2

一组彩色、风格化的图标，具有柔和色调现实主义风景风格，浅粉色和浅琥珀色的，社交媒体艺术，M42 底座，禄来福来原创，弗拉基米尔·库什，浅白色和浅灰色 --ar 1：1 --q 2

关键词组合

a set of blue ui kits for apps, in the style of photorealistic detailing, gradient color blends, hendrick cornelisz vroom, realistic detailing, matte background, soft color fields --ar 3:4

一组用于应用程序的蓝色 UI 套件，采用逼真的细节风格，渐变颜色混合，亨德里克·科涅里茨·弗鲁，逼真的细节，亚光背景，柔和的色域 --ar 3：4

关键词组合

an apple watch with all kinds of light effects on it, in the style of cyberpunk realism, highly detailed illustrations, steampunk influences, dark cyan and amber, precision engineering, dynamic energy flow --ar 1:1

带有各种光效的苹果手表，采用赛博朋克现实主义风格，高度详细的插图，蒸汽朋克影响，深青色和琥珀色，精密工程，动态能量流 --ar 1：1

关键词组合

the apple watch uses various colorful bubbles of the day, in the style of naoto hattori, luminous palette, uhd image, abstract illusionism, dark amber, detailed background elements, avant-garde design --ar 3:4

苹果手表采用了当时的各种七彩泡泡，具有服部直人风格，夜光调色板，超高清图像，抽象幻觉，深琥珀色，细节背景元素，前卫设计 --ar 3：4

关键词组合

an image for the website with the name zuziize suzete, in the style of flat illustrations, rusticcore, golden hues, installation-based, photorealistic pastiche, website, dau al set --ar 1:1

名为 zuziize suzete 的网站图片，采用平面插图风格，乡村风格，金色色调，基于装置，照片级逼真拼贴，网站，达乌塞 --ar 1 : 1

关键词组合

a blue website design for a professional business, in the style of dark silver and aquamarine, ilya kuvshinov, detailed facial features, conceptual themes, intensely detailed, retro, fine detailed --ar 1:1

为专业企业设计的蓝色网站，采用深银色和海蓝色风格，伊利亚·库夫希诺夫，详细的面部特征，概念主题，强烈的细节，复古，精细的细节 --ar 1 : 1

关键词组合

an abstract colorful ui ui layout with videos and apps, in the style of organic fluid shapes, jan pietersz saenredam, cyril rolando, vibrant academia, eye-catching, gustave van de woestijne, realistic color schemes --ar 1:1

带有视频和应用程序的抽象多彩 UI 布局，具有有机流体形状的风格，扬·彼得斯·萨恩雷丹，西里尔·罗兰多，充满活力的学术界，引人注目，古斯塔夫·凡·德·沃斯提因，逼真的配色方案 --ar 1：1

关键词组合

web design for bakery web shop ppse, in the style of eiko ojala, gold and amber, tivadar csontváry kosztka, realistic yet stylized, romantic gestures, oil painter, storybook-like --ar 1:1

面包店网店 ppse 的网页设计，采用伊卡·奥贾拉风格，金色和琥珀色，蒂华达·贡茨卡·贡特瓦里，逼真但程式化，浪漫的姿态，油画，故事书般的风格 --ar 1：1

关键词组合

in the style of cyril rolando, dark cyan, innovative page design, geometric animal figures, organic and flowing forms, grandiose color schemes, luminous 3d objects --ar 3:4

西里尔·罗兰多的风格，深青色，创新的页面设计，几何动物图形，有机和流动的形式，宏伟的配色方案，发光的3D物体 --ar 3：4

关键词组合

the website has a colorful design in 3d, in the style of hendrik weissenbruch, peter mohrbacher, webcore, light black and purple, colorful dreams, matte background, birds-eye-view --ar 16:9

该网站有一个彩色的 3D 设计，采用亨德里克·魏森布鲁赫的风格，彼得·莫尔巴赫，网络风，浅黑色和紫色，彩色梦境，亚光背景，鸟瞰图 --ar 16：9

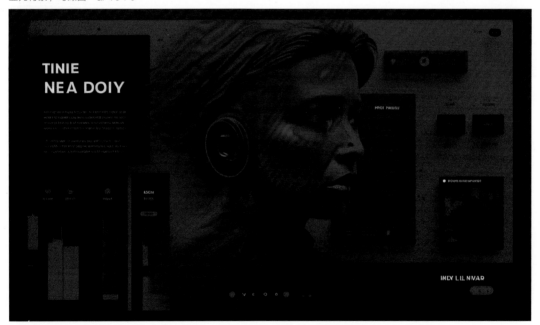

关键词组合

the red and black ui design created by sanag, in the style of monochromatic artworks, isometric, matte photo, organic fluidity, neo-academism, color splash, use of screen tones --ar 16:9

塞那创作的红黑 UI 设计，采用单色艺术品风格，等距，亚光照片，有机流动性，新学院主义，色彩飞溅，使用屏幕色调 --ar 16：9

关键词组合

a website for a gaming company, in the style of mythological realism, textured illustrations, argus c3, matte background, yup'ik art, fine detailed, soft color fields --ar 3:4

一家游戏公司的网站，采用神话现实主义风格，纹理插图，阿格斯 C3，亚光背景，尤皮克艺术，精细的细节，柔和的色域 --ar 3：4

关键词组合

page web and web app design layouts, in the style of bold spray-painted letters, fluid formations, isometric, indonesian art, an energetic and chaotic style --ar 16:9

页面网页和网页应用程序设计布局，采用大胆的喷漆字母风格，流体构造，等距，印度尼西亚艺术，充满活力和混乱的风格 --ar 16：9

关键词组合

a colorful layout for a man in glasses, in the style of colorful gradients, interactive media, dark and brooding designer, 3d, vibrant color fields, organic and fluid, germanic art --ar 16:9

为戴眼镜的男人设计的彩色布局，采用彩色渐变风格，交互式媒体，黑暗和沉思的设计师，3D，充满活力的色域，有机且流畅，日耳曼艺术 --ar 16：9

3.7 服装类

关键词组合

upscaling image #2 with the royal garb in blue and gold, in the style of han dynasty, realist fine details, dark turquoise and light crimson, artifacts of online culture, textured detail, historical documentation, karencore --ar 3:4 --v 5

放大图片 #2，蓝色和金色的皇家服装，汉代风格，现实主义的精细细节，深绿松石色和浅绛色，网络文化的文物，纹理细节，历史文献，克伦风 --ar 3：4 --v 5

关键词组合

oriental style dress for a woman, in the style of meticulous photorealistic still lifes, japanese-inspired imagery, light gray and orange, hannah flowers, concept art, dark azure and white, exquisite clothing detail --ar 3:4 --v 5

东方风格的女性连衣裙，采用细致的摄影静物风格，日本灵感的图像，浅灰色和橙色，汉娜花，概念艺术，深蓝色和白色，精致的服装细节 --ar 3：4 --v 5

关键词组合

with a thigh slit, in the style of asymmetrical geometry, flowing draperies, whimsical feel, bold yet graceful, light beige, charly amani, classical style --ar 3:4 --v 5

大腿开叉的姿势，采用不对称的几何风格，飘动的布料，异想天开的感觉，大胆而优雅，浅米色，查理·阿曼尼，古典风格 --ar 3：4 --v 5

关键词组合

a man wearing a cape on his torso, in the style of jean nouvel, japanese-inspired, igor kieryluk, traditional techniques reimagined, chris cunningham, dau al set, pop-culture-infused --ar 1:1 --q 2 --v 5

一个穿着斗篷的男人，具有让·努维尔的风格，受到日本的启发，伊戈尔·基鲁克，传统技术的重新想象，克里斯·康宁汉，达乌塞，流行文化的融合 --ar 1：1 --q 2 --v 5

关键词组合

fashion design by sigma concept fashions, in the style of stark visuals, light pink and silver, the stars art group (xing xing), slumped/draped, pop inspo, luxurious drapery, realistic chiaroscuro --ar 1:1 --q 2 --v 5

出自西格玛概念时装公司的时尚设计，以鲜明的视觉效果为风格，浅粉色和银色，星星艺术团（兴兴），倾斜 / 悬垂，流行的灵感，奢华的布料，逼真的明暗对照 --ar 1：1 --q 2 --v 5

关键词组合

a woman wearing a colorful coat on the runway, in the style of avant-garde ceramics, he jiaying, gray and azure, zuckerpunk, johnson tsang, pure color, haifa zangana --q 2 --v 5

一个女人在 T 台上穿着五颜六色的大衣，具有前卫陶瓷的风格，何家英，灰色和天蓝色，扎克朋克，曾章成，纯色，海法·扎加娜 --q 2 --v 5

关键词组合

model dressed washed denim suit with wrinkles and holes and snags and inside of shirt with slanted navy collar, asymmetric structure, deconstruction, distorted silhouettes, full body views, jeans and cotton fabric, paris fashion show, realistic photography, spotlights, hd, 8k --ar 1:1 --q 2 --v 5

模特穿着水洗牛仔套装，有褶皱、破洞和卡扣，衬衫内部有倾斜的海军领，不对称结构，解构主义，扭曲的轮廓，全身视图，牛仔裤和棉织物，巴黎时装秀，写实摄影，聚光灯，高清，8K--ar 1:1 --q 2 --v 5

3.8 创意类

关键词组合

man writing with paper on top of head, in the style of photorealistic fantasies, captivating lighting, mind-bending sculptures, bulbous, textured illustrations, bold and busy, tabletop photography --ar 3:4 --v 5

男人在面前的纸上写字，具有真实感的幻想风格，迷人的灯光，令人费解的雕塑，球根状，有纹理的插图，大胆而忙碌，桌面摄影 --ar 3：4 --v 5

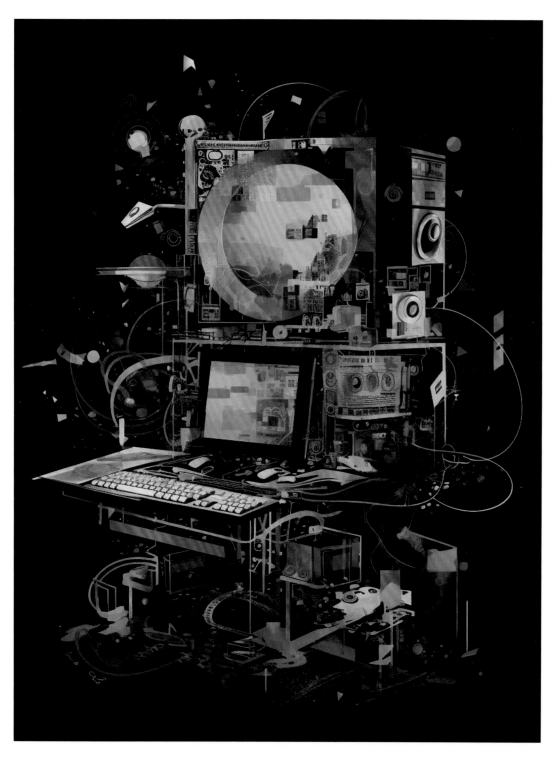

关键词组合

a computer desktop with a lot of colors and symbols, in the style of mechanical realism, wlop, mordecai ardon, concert poster, optical illusionism, webcore, neo-dadaism --ar 3:4 --v 5

一个有很多颜色和符号的电脑桌面，采用机械现实主义风格，王凌，莫迪凯·阿尔顿，音乐会海报，视觉错觉，网络风，新达达主义 --ar 3：4 --v 5

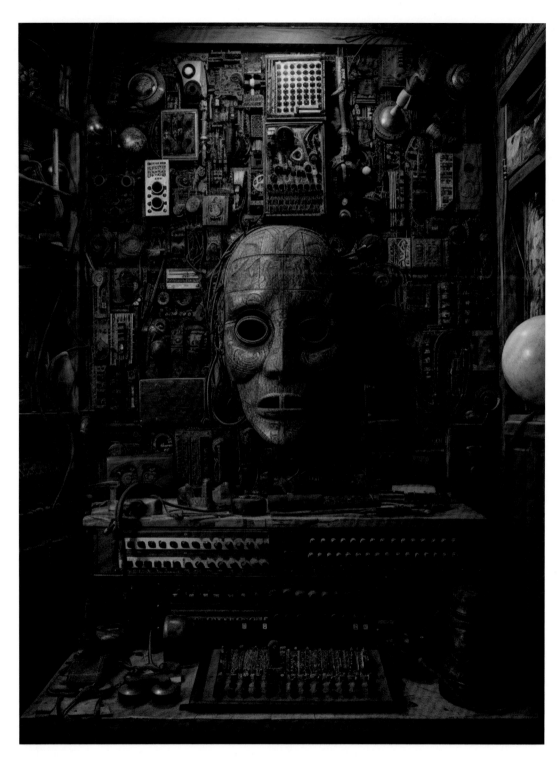

关键词组合

modular equipment and modular synthesizer, in the style of anton semenov, matthias haker, uhd image, iconic, georges de la tour, studio light, masks and totems --ar 3:4 --v 5

模块化设备和模块化合成器，采用安东·赛梅诺夫风格，马蒂亚斯·哈克，超高清图像，符号，乔治·德·拉·图尔，工作室灯光，面具和图腾 --ar 3：4 --v 5

关键词组合

ceramic guns with guns on a table and some blue vase, in the style of photographically detailed portraitures, digitally manipulated, rural china, nature-inspired camouflage, duckcore, canon 7, detailed ink --ar 16:9 --v 5

桌上放着陶瓷枪和一些蓝色花瓶，采用详细的摄影肖像画风格，数字处理，中国乡村，自然迷彩，鸭核，佳能7，细致的油墨 --ar 16：9 --v 5

关键词组合

a man standing in a 3d image standing on top of a tree, in the style of surrealistic portraits, indian pop culture, mashup of styles, wildlife photography, cabincore, indonesian art, stock photo --ar 16:9 --v 5

一个人站在3D图像中，站在树顶，超现实主义肖像风格，印度流行文化，风格混搭，野生动物摄影，木屋美学，印度流行艺术，库存照片 --ar 16：9 --v 5

关键词组合

a woman in a tall glass dome looking at a palm tree, in the style of retro-futuristic cyberpunk, elaborate landscapes, afrofuturism- inspired, matte painting, hyper-detailed, shot on 70mm --ar 16:9 --v 5

一个女人在高大的玻璃穹顶下看着一棵棕榈树，复古未来主义赛博朋克风格，精致的风景，受非洲未来主义启发，亚光绘画，超详细，使用 70mm 镜头拍摄 --ar 16：9 --v 5

关键词组合

a made of rocks for a woman, in the style of japenese renaissance, realistic yet imaginative, intricate costumes, mixed-media sculptor, dark beige and azure, heian period, layered fibers --ar 3:4 --v 5

一个用石头做成的女人，日本文艺复兴风格，逼真而富有想象力，复杂的服装，混合媒体雕塑家，深米色和天蓝色，平安时代，分层纤维 --ar 3：4 --v 5

关键词组合

an image in which boba fett is holding flowers in his armor, in the style of orange and gold, eve ventrue, gloomy, sharp brushwork, alois arnegger --ar 3:4 --v 5

波巴·费特身着盔甲，手持花朵，橙色和金色风格，夏娃·文特鲁，阴沉，锐利的笔触，阿洛伊斯·阿内格 --ar 3：4 --v 5

关键词组合

a cat dressed as a pirate is standing on a boat, in the style of realistic and hyper-detailed renderings, cryengine, photo-realistic, 32k uhd, realistic animal portraits, close up, atmospheric scenes --ar 3:4 --v 5

一只打扮成海盗的猫站在船上，逼真超精细渲染风格，悲伤王子引擎，逼真，32K 超高清，逼真动物肖像，特写，大气场景 --ar 3：4 --v 5

关键词组合

a gold female character standing in space, in the style of photorealistic detailing, chaotic energy, light white and light gold, realistic hyper-detail, detailed facial features, body extensions --ar 3:4 --v 5

站在太空中的金色女性角色，逼真的细节风格，混乱的能量，浅白色和浅金色，逼真详细的面部特征，身体延伸 --ar 3：4 --v 5

关键词组合

neon zebra artwork, in the style of realistic hyper-detailed rendering, flat chromatic fields, hyper-realistic animal illustrations, redshift, flat colors, distinctive noses, post processing --ar 16:9 --v 5

霓虹灯斑马艺术作品，以逼真的超详细渲染风格，平坦的色域，超逼真的动物插图，红移，颜色平淡，独特的鼻子，后期处理 --ar 16：9 --v 5

关键词组合

portrait of a koi in a pond with roses, in the style of hyper-realistic animal illustrations, dark gold and light crimson, fantastical creatures, intense close-ups, airbrushing, light teal and dark crimson, depictions of animals --ar 16:9 --v 5

玫瑰池塘中锦鲤的肖像，采用超写实动物插图风格，深金色和浅绛色，奇幻生物，强烈的特写镜头，喷绘，浅蓝绿色和深红色，动物描绘 --ar 16：9 --v 5

关键词组合

a collection of items and parts for an iron man suit, in the style of 8k resolution, marcin sobas, 32k uhd, john larriva, use of impasto technique, limited color range --ar 16:9 --v 5

钢铁侠套装的物品和零件集合，8K 分辨率风格，马尔钦·索巴斯，32K 超高清，约翰·拉里瓦，使用厚涂技术，有限的颜色范围 --ar 16：9 --v 5

关键词组合

a beautiful bird with gold feathers and colorful feathers sitting in a tree, in the style of hyper-realistic sci-fi, detailed character illustrations, eugene delacroix, glowing colors, layers of texture, enigmatic tropics --ar 16:9 --v 5

一只美丽的鸟，有金色的羽毛和彩色的羽毛，落在树上，超现实的科幻风格，详细的人物插图，欧仁·德拉克罗瓦，发光的颜色，有层次，神秘的热带 --ar 16：9 --v 5

关键词组合

an aerial view of a modern futuristic space ship, in the style of classical symmetry, dark white and amber, photographically detailed portraitures, cabincore, handheld, grandeur of scale --ar 3:4 --v 5

现代未来主义太空飞船的鸟瞰图，采用经典的对称风格，深白色和琥珀色，摄影细节肖像，舱内，手持，规模宏大 --ar 3：4 --v 5

关键词组合

a huge dragon standing on top of a rock in a field of fire at sunset, in the style of gothic grandeur, dark orange and gold, rich and immersive, realistic cityscapes, ashcan school, uhd image, calculated --ar 3:4 --v 5

日落时分，一条巨大的龙站在火场中的岩石顶部，具有哥特式的宏伟风格，深橙色和金色，丰富而身临其境，逼真的城市景观，垃圾箱画派，超高清图像，计算 --ar 3：4 --v 5

关键词组合

painting of a lady surrounded by birds, in the style of brian despain, mark ryden, dark white and dark aquamarine, victorian-era clothing, realistic details, pigeoncore --ar 16:9 --v 5

一幅画有被鸟包围的女士的画作，布瑞恩·德斯潘风格，马克·莱登，深白色和深海蓝宝石色，维多利亚时代的服装，逼真的细节，鸽子 --ar 16：9 --v 5

关键词组合

star wars the force awakens shadow of the sabernacle character wallpaper, in the style of lovecraftian, realistic detail, janek sedlar, heavily textured, ultra hd --ar 16:9 --v 5

《星球大战原力觉醒》剑影人物壁纸，洛夫克拉夫特风格，逼真的细节，雅内克·塞德拉，重纹理，超高清 --ar 16：9 --v 5

关键词组合

animated bird on dark background, in the style of intricately sculpted, colorful storytelling, rococo-inspired art, zbrush, polychrome terracotta, pigeoncore, intricate illustrations --ar 16:9 --v 5

深色背景上栩栩如生的鸟，复杂的雕刻风格，丰富多彩的故事，洛可可风格的艺术、ZBrush、彩色陶瓦，鸽子，复杂的插图 --ar 16：9 --v 5

关键词组合

the image features black skull balloons, in the style of dark tones, photo taken with nikon d750, contest winner, fairycore, edgy --ar 16:9 --v 5

该图片以黑色骷髅气球为特色，采用暗色调风格，用尼康 D750 拍摄的照片，竞赛获胜者，精灵风，前卫的 --ar 16：9 --v 5

关键词组合

a skull is set with colorful objects, in the style of realistic and hyper-detailed renderings, floral explosions, focus stacking, kimoicore, dark and brooding designer, realistic hyper-detail, foampunk --ar 3:4 --v 5

头骨上镶嵌着五颜六色的物体，采用逼真的超细节渲染风格，花卉爆炸，焦点堆叠，让人不适的风格，黑暗和沉思的设计师，逼真的超细节，泡沫朋克 --ar 3：4 --v 5

关键词组合

the egyptian pharaoh, a beautiful and regal painting, in the style of cyberpunk imagery, manticore, bronze and blue, hyper-realistic animal illustrations, comic art, d&d, junglepunk --ar 3:4 --v 5

埃及法老，一幅美丽而富丽堂皇的画作，采用赛博朋克意象，蝎狮，青铜色和蓝色，超写实动物插图，漫画艺术，《龙与地下城》，丛林朋克 --ar 3：4 --v 5

3.9 影视类

关键词组合

pacific rim poster featuring big robots and people standing by, in the style of orange and azure, post-apocalyptic backdrops, uhd image, brutal action --ar 3:4 --v 5 --q 2

以大型机器人和站在一旁的人为特色的《环太平洋》海报，橙色和天蓝色风格，后世界末日背景，超高清图像，残酷动作 --ar 3：4 --v 5 --q 2

关键词组合

guardians of the galaxy 2 movie poster, in the style of bold, saturation innovator, 8k resolution, rendered in maya, imax, distinctive noses, violet and aquamarine, glowwave --ar 3:4 --v 5

《银河护卫队 2》电影海报，采用大胆的饱和度，创新者的风格，8K 分辨率，在 Maya 中渲染，IMAX，独特的鼻子，紫罗兰色和海蓝宝石，辉光波 --ar 3：4 --v 5

关键词组合

a poster for the chinese film called big x, in the style of artificial environments, apocalypse art, ocean academia, ethereal sculptures, burned/charred, light green and blue, nul group --ar 3:4 --v 5

中国电影《Big X》的海报，人工环境风格，末日艺术，海洋学术，空灵雕塑，燃烧 / 烧焦，浅绿色和蓝色，NUL group--ar 3：4 --v 5

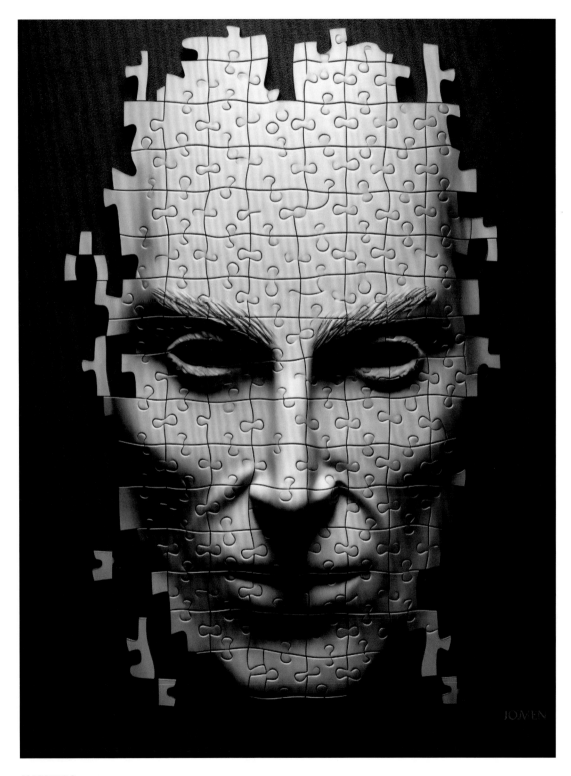

关键词组合

an image of the movie poster for jigsaw, in the style of leica cl, human-canvas integration, classical portraiture, carving, sharp and angular --ar 3:4

拼图电影海报图片，徕卡 CL 风格，人画融合，古典人像，雕刻，棱角分明 --ar 3：4

关键词组合

an imposing figurine of an animal dressed in skeletons, in the style of fractal geometry, light gray and gold, hyperrealistic murals, dragoncore, organic architecture, symmetrical chaos --ar 3:4 --v 5

一个气势恢宏的动物骷髅雕像，分形几何风格，浅灰色和金色，超现实主义壁画，龙核，有机建筑，混乱的对称 --ar 3：4 --v 5

关键词组合

a poster with two men on a blue car, in the style of isaac julien, zack snyder, zhang kechun, nostalgic, pure color, tightly composed scenes, ferrania p30 --ar 3:4

两个男人在蓝色汽车上的海报，艾萨克·朱利安、扎克·施奈德，张克纯的风格，怀旧，纯色，紧凑的场景，菲拉尼亚 P30 --ar 3：4

关键词组合

a poster of warped remastered from the movie, ranked 12, dark sky-blue and orange, futuristic robots, gabriel bá, dark silver and gold, animal intensity, craft --ar 3:4 --q 2

根据电影重新制作的海报，排名第 12，深天蓝色和橙色，未来机器人，加布里埃尔·巴，深银色和金色，动物强度，工艺 --ar 3：4 --q 2

关键词组合

the poster for the third video game, in the style of realistic fantasy artwork, detailed facial features, dark green and gold, threaded tapestries --ar 3:4 --q 2

第三个视频游戏的海报，采用逼真的幻想艺术风格，详细的面部特征，深绿色和金色，螺纹挂毯 --ar 3：4 --q 2

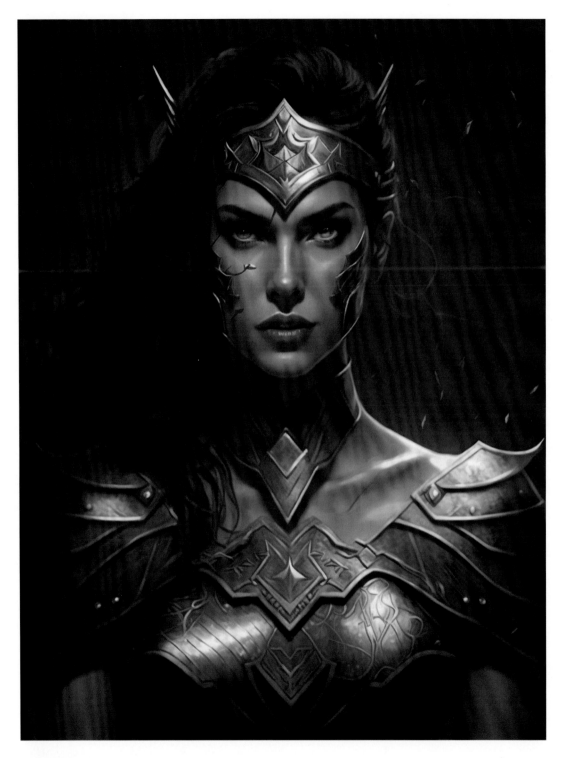

关键词组合

an illustration of a woman in wonder woman armor, in the style of realistic portrait painter, dark bronze and light bronze, princesscore, indian pop culture, celebrity and pop culture references, dignified poses, eerily realistic --ar 3:4

一幅穿着神奇女侠盔甲的女性插图，采用写实肖像画家的风格，深青铜色和浅青铜色，王室风，印度流行文化，名人和流行文化参考，端庄的姿势，怪诞的现实主义 --ar 3：4

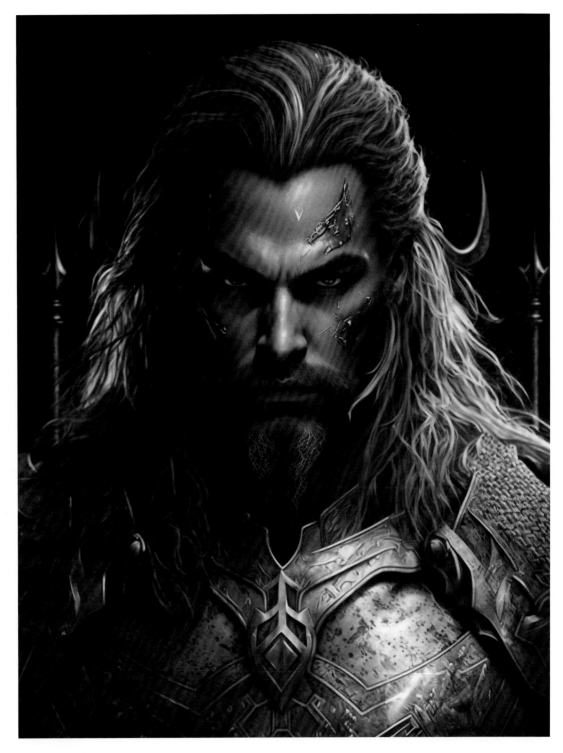

关键词组合

justice league aquaman wallpapers, in the style of realistic and hyper-detailed renderings, dark gold and white, medieval fantasy, hyper-realistic portraiture, mountainous vistas, neo-geo, realistic hyper-detail --ar 3:4

《正义联盟》的海王壁纸，具有逼真和超细节渲染的风格，暗金色和白色，中世纪幻想，超逼真的肖像画，山景，新地理，逼真的超细节 --ar 3：4

关键词组合

an epic art of the incredible hulk, in the style of aggressive digital illustration, full body --ar 3:4

《无敌浩克》的史诗艺术，采用激进的数字插图风格，全身 --ar 3：4

关键词组合

dreamgate, surrealism, mad science and the surreal with all the latest info, in the style of pensive surrealism, neo-classical symmetry, hyper-realistic water, vintage imagery, subdued tranquility, classical landscapes, frontal perspective --ar 16:9 --v 5

梦幻之门，超现实主义，疯狂科学和超现实主义，提供所有最新信息，沉思的超现实主义风格，新古典对称，超现实主义的水，复古意象，柔和的宁静，古典风景，正面透视 --ar 16：9 --v 5

关键词组合

2070 film still of the, in the dark and damp basement, a middle-aged man was holding a gem in his hand,8k,enhancing science fiction technology elements, enhanced metatemporal elements, mechanical elements,ultra clear realism --ar 16:9 --v 5

《全面回忆 2070》电影剧照，阴暗潮湿的地下室里，一个中年男人手里拿着一颗宝石，8K，增强科幻科技元素，增强元时空元素，机械元素，超清写实 --ar 16：9 --v 5

关键词组合

images of a scene of several shots of a scene in shinobi, in the style of ink wash painter, blown-off-roof perspective, medieval inspiration, dynamic and action-packed scenes, chinapunk, ricoh ff-9d, mixed media marvel --ar 16:9 --v 5

《甲贺忍法帖》中几个镜头的场景图像，水墨画风格，屋顶视角，中世纪灵感，动态和动感十足的场景，中国朋克，理光 FF-9D，复合媒介漫威 -- ar 16：9 --v 5

关键词组合

black screens with the silhouettes of people and cars, accompanied by cars in the background, in the style of realistic landscapes with soft, tonal colors, emotionally charged scenes, narrative-driven visual storytelling, fisheye effects, multi-panel compositions, dark, muted colors, eerily realistic --ar 16:9 --v 5

黑色屏幕上有人和车的剪影，背景是车，具有柔和色调的逼真景观风格，充满情感的场景，叙事驱动的视觉故事，鱼眼效果，多面板结构，深色，柔和的颜色，诡异又逼真 --ar 16：9 --v 5

关键词组合

a page with illustrations of an urban scene, in the style of cinematic lighting, dark palette, daniel f. gerhartz, monochrome canvases, cartoon compositions, maria kreyn, cliff chiang --ar 16:9 --v 5

带有城市场景插图的页面，采用电影照明风格，深色调色板，丹尼尔·格哈茨，单色画布，卡通作品，玛利亚·克瑞恩，克里夫·江 --ar 16：9 --v 5

关键词组合

in the style of photorealistic urban scenes, ambient occlusion, leica cl, depictions of theater, post- war french design, realistic and hyper-detailed renderings --ar 16:9 --v 5 --q 2

风格逼真的城市场景，环境遮挡，徕卡 CL，剧院描绘，战后法国设计，逼真且超详细的渲染 --ar 16 : 9 --v 5 --q 2

关键词组合

in the style of film noir style, multiple screens, highly detailed illustrations, dark white and dark pink, matte photo, high-contrast shading, innovative page design --ar 16:9 --v 5 --q 2

黑色电影风格，多屏幕，高度详细的插图，深白色和深粉色，亚光照片，高对比度的阴影，创新的页面设计 --ar 16 : 9 --v 5 --q 2

关键词组合

one of three pictures of children from beethoven being shown with pianos and statues, in the style of stop-motion animation, japanese photography, portraits with soft lighting, uhd image, collage-based, classical, historical genre scenes, video --ar 16:9 --v 5

贝多芬的孩子与钢琴和雕像一起展示的三幅照片之一，定格动画风格，日本摄影，柔和灯光肖像，超高清图像，拼贴画，古典，历史流派场景，视频 --ar 16：9 --v 5

关键词组合

a man's face, with passengers on a train, with a bus and a train full of men, in the style of cinematic stills, japanese inspiration, massurrealism, grandparentcore, melancholic cityscapes, multiple screens, candid shots of famous figures --ar 16:9 --v 5

一个男人的脸，火车上的乘客，一辆公共汽车和一列满载男人的火车，电影剧照风格，日本灵感，大众超现实主义，祖父母风格，忧郁的城市景观，多个屏幕，著名人物的快照 --ar 16：9 --v 5

ART
APPRECIATION

艺术赏析

列奥纳多·达·芬奇
Leonardo da Vinci

意大利文艺复兴时期画家、自然科学家、工程师，与米开朗琪罗、拉斐尔并称"文艺复兴后三杰"。达·芬奇思想深邃，学识渊博，擅长绘画、雕刻、发明、土木，通晓数学、生物学、物理学、天文学、地质学等学科，是人类历史上少见的全才。

巴勃罗·毕加索
Pablo Picasso

西班牙画家、雕塑家，是现代艺术的创始人，西方现代派绘画的主要代表。毕加索是当代西方最有创造性和影响最深远的艺术家，是 20 世纪最伟大的艺术天才之一。

文森特·凡·高
Vincent van Gogh

荷兰后印象派画家。代表作有《星月夜》、自画像系列、向日葵系列等。
他早期画风写实，受到荷兰传统绘画及法国写实主义画派的影响。1886 年，
他来到巴黎，结识印象派和新印象派画家，并接触到日本浮世绘的作品，
视野的扩展使其画风巨变。

奥斯卡 - 克劳德·莫奈
Oscar-Claude Monet

法国画家，被誉为"印象派领导者"，是印象派代表人物和创始人之一。莫奈擅长光与影的实验与表现技法。他最明显的风格是改变了阴影和轮廓线的画法，在莫奈的画作中看不到非常明确的阴影，也看不到突显或平涂式的轮廓线。

保罗 · 塞尚
Paul Cézanne

法国后印象主义画派画家，他的作品和理念影响了 20 世纪许多艺术家和艺术运动，尤其是立体派。塞尚的最大成就是对色彩与明暗进行了前所未有的精辟分析，颠覆了以往的视觉透视点，这是以往任何绘画流派都无法做到的。他被誉为"现代艺术之父"。

亨利·马蒂斯
Henri Matisse

法国著名画家、雕塑家、版画家，野兽派创始人和主要代表人物，代表作有《奢华、宁静与愉快》《生活的欢乐》《开着的窗户》《戴帽的妇人》等。他以使用鲜明大胆的色彩而闻名。

伊凡·伊凡诺维奇·希施金
Ivan I. Shishkin

杰出的俄罗斯风景画家，是 19 世纪下半叶现实主义风景画的奠基人之一。巡回展览画派的创始人，彼得堡美术院院士与教授，他与列维坦、艾瓦佐夫斯基和库因哲等人都是当时俄罗斯著名的风景画家。

伊萨克·列维坦
Levitan Isaak Iliich

俄罗斯杰出的写生画家，现实主义风景画大师，巡回展览画派的成员之一。列维坦的作品极富诗意，深刻而真实地表现了俄罗斯大自然的优美风景与多方面的特点。

伊里亚·叶菲莫维奇·列宾
Ilya Efimovich Repin

俄罗斯杰出的批判现实主义画家，巡回展览画派重要代表人物。先后旅居意大利及法国，研究欧洲古典及近代美术。而后创作了大量的历史画、风俗画和肖像画。

保罗·高更
Paul Gauguin

法国后印象派画家、雕塑家，代表作品有《我们从何处来？我们是谁？我们向何处去？》《黄色基督》《游魂》《敬神节》等。保罗·高更与凡·高、塞尚并称为"后印象派三大巨匠"。

伦勃朗·哈尔曼松·凡·莱因
Rembrandt Harmenszoon
van Rijn

欧洲 17 世纪最伟大的画家之一，也是荷兰历史上最伟大的画家。伦勃朗早年师从 P. 拉斯特曼，1625 年在家乡开设画室。画作体裁广泛，擅长肖像画、风景画、风俗画、宗教画、历史画等领域。

拉斐尔·桑西
Raffaello Santi

意大利著名画家，也是"文艺复兴后三杰"中最年轻的一位，代表了文艺复兴时期艺术家从事理想美的事业所能达到的巅峰。他的性情平和、文雅，创作了大量的圣母像，他的作品充分体现了安宁、协调、和谐、对称，以及完美和恬静的秩序。

米开朗琪罗·博那罗蒂

Michelangelo Buonarroti

意大利文艺复兴时期伟大的绘画家、雕塑家、建筑师和诗人，文艺复兴时期雕塑艺术最高峰的代表，与拉斐尔·桑西和达·芬奇并称为"文艺复兴艺术三杰"。米开朗琪罗的代表作有《大卫》《创世纪》等。

瓦西里·康定斯基
Wassily Kandinsky

画家和美术理论家。现代
艺术的伟大人物之一，同
时也是现代抽象艺术在理
论和实践上的奠基人。

徐悲鸿
Xu Beihong

中国现代画家、美术教育家，景星学社社员。曾留学法国学西画，归国后长期从事美术教育，先后任教于国立中央大学艺术系、北平大学艺术学院和北平艺专，1949 年任中央美术学院院长。主要绘画作品有《奔马图》《愚公移山图》《田横五百士》等。

齐白石
Qi Baishi

中国近现代书画家、书法篆刻家。主要绘画作品有《墨虾》《牧牛图》《蛙声十里出山泉》等。

KEYWORD
INSPIRATION

关键词灵感

与人物有关的	
女人	woman
女孩	girl
男人	man
男孩	boy
奶奶	grandmother
爷爷	grandfather
爸爸	father
妈妈	mother
叔叔	uncle
阿姨	aunt
外婆	grandmother
外公	grandfather
兄弟	brother
姐妹	sister
朋友	friend
同事	colleague
邻居	neighbor
学生	student
父母	parents
教师	teacher
工人	worker
记者	reporter
演员	actor
厨师	chef
医生	doctor
护士	nurse
司机	driver
军人	soldier
律师	lawyer
商人	merchant
会计师	accountant
店员	clerk
出纳员	cashier
作家	writer
导游	guide
模特	model
警察	police
歌手	singer
画家	painter
裁缝	tailor
翻译	translator
法官	judge
保安	security guard
花匠	gardener

与人物有关的	
服务员	waiter
清洁工	cleaner
建筑师	architect
理发师	barber
采购员	buyer
设计师	designer
消防员	fireman
推销员	salesman
魔术师	magician
邮递员	postman
售货员	salesperson
救生员	lifeguard
运动员	athlete
工程师	engineer
飞行员	pilot
管理员	administrator
机械师	mechanic
经纪人	broker
审计员	auditor
漫画家	cartoonist
科学家	scientist
主持人	host
调酒师	bartender
化妆师	dresser
音乐家	musician
艺术家	artist
糕点师	baker
甜品师	dessert chef
外交官	diplomat
舞蹈家	dancer
弓箭手	archer
钢琴家	pianist
酒吧老板	bar owner
机长	captain
空姐	airline stewardess
赛车手	racing driver
特警	special police
私人侦探	private detective
探险家	explorer
盗贼	thief
赛博格	cyborg
黑客	hacker
警察	police
机械工程师	mechanical engineer

与人物有关的

特种兵	special operator
赛博格战士	cyborg warrior
超级英雄	superhero
雇佣兵	mercenary
独立记者	independent journalist
AI 程序员	ai programmer
发明家	inventor
太空探险家	space explorer
未来艺术家	future artist
无业游民	unemployed person
退休人员	retiree
自由职业者	freelancer
专业投资人	professional investor
公务员	civil servant
政治家	politician
奥林匹克运动员	olympic athlete
游泳运动员	swimmer
足球运动员	soccer player
健身教练	fitness instructor
瑜伽师	yogi
保姆	babysitter
家政服务人员	domestic helper
电工	electrician
管道工	plumber
顾问	consultant
经理	manager
警官学员	cadet
外卖骑手	takeaway rider
快递员	courier
上班族	commuter

动物

大熊猫	panda
黑熊	black bear
鹿	deer
树懒	sloth
貂	mink
斑马	zebra
狐	fox
豹子	leopard
麝牛	musk ox
狮子	lion
考拉	koala

动物

长颈鹿	giraffe
猩猩	orangutan
海豚	dolphin
鸭嘴兽	platypus
北极熊	polar bear
袋鼠	kangaroo
河马	hippopotamus
鲸鱼	whale
雪豹	snow leopard
金丝猴	golden monkey
梅花鹿	sika deer
鸟	bird
蝴蝶	butterfly
蜻蜓	dragonfly
恐龙	dinosaur
驴	donkey
羊驼	alpaca
猫	cat
鹰	eagle
鹅	goose
鸡	chicken
鸭	duck
鸽子	pigeon
啄木鸟	woodpecker
鹦鹉	parrot
杜鹃鸟	cuckoo
鸵鸟	ostrich
喜鹊	magpie
燕子	swallow
夜莺	nightingale
金丝雀	canary
画眉鸟	throstle
企鹅	penguin
变色龙	chameleon
青蛙	frog
蜜蜂	honeybee
蜘蛛	spider
萤火虫	firefly
蝉	cicada
瓢虫	ladybug
鱿鱼	squid
虾	shrimp
龙虾	lobster
螃蟹	crab

动物	
孔雀	peacock
海龟	sea turtle
蚂蚁	ant
鱼	fish
大象	elephant
海豹	seal
鹤	crane
鲨鱼	shark
海星	starfish
海螺	conch
海草	seaweed
扇贝	scallop
水母	jellyfish
猛犸象	mammoth
珊瑚	coral
海参	sea cucumber
海鸥	seagull
灯笼鱼	lantern fish
海燕	petrel
公鸡	rooster
母鸡	hen
小鸡	chick

植物	
花	flower
草	grass
树	tree
银杏树	ginkgo tree
杨树	poplar
桃树	peach tree
槐树	locust tree
柳树	willow
松树	pine
枫树	maple
杨树	poplar
红木	mahogany
橡胶树	rubber tree
椰子树	coconut tree
枫叶	maple leaf
红叶	red leaf
小麦	wheat
稻子	rice
玉米	corn

植物	
葵花	sunflower
丁香花	lilac
桂花	osmanthus fragrans
山茶花	camellia
桑树	mulberry
多肉植物	succulent plant
仙人球	prickly pear
迷迭香	rosemary
文竹	asparagus fern
绿萝	epipremnum aureum
芦荟	aloe
人参	ginseng
铁树	iron tree
虎尾兰	sansevieria
荷花	lotus
月季	chinese rose
玫瑰	rose
杜鹃花	rhododendron
牵牛花	morning glory
牡丹花	peony
樱花	cherry blossoms
仙人掌	cactus
郁金香	tulip
百合花	lily
薰衣草	lavender
迎春花	jasminum nudiflorum
玉兰花	magnolia flower
紫罗兰	violet
水仙	narcissus
菊花	chrysanthemum
含笑花	banana shrub
睡莲	water lily
萝卜	radish
茄子	eggplant
黄瓜	cucumber
大白菜	chinese cabbage
菠菜	spinach
豆芽	bean sprout
生菜	lettuce
青椒	green pepper
芹菜	celery
胡萝卜	carrot
洋葱	onion
姜	ginger

植物	
蒜	garlic
西红柿	tomato
豆角	long bean
土豆	potato
苹果	apple
西瓜	watermelon
杧果	mango
火龙果	pitaya
葡萄	grape
梨	pear
橘子	tangerine
橙子	orange
榴莲	durian
草莓	strawberry
香蕉	banana
桃子	peach
柠檬	lemon
菠萝蜜	jackfruit
猕猴桃	kiwi fruit
圣女果	cherry tomato

发型	
短发	short hair
直发	straight hair
卷发	curly hair
波浪卷发	wavy hair
长发	long hair
半长发	semi-long hair
齐刘海长发	long hair with bangs
中分卷发	mid parted curly hair
不规则短发	irregular short hair
高马尾	high ponytail
莫西干发型	mohican
平头	crew cut
子弹头发型	bullet haircut
大背头	slicked back hair
朋克发型	punk haircut
斜庞克发型	oblique punk haircut
纹理烫	textured perm

脸型	
瓜子脸	oval face
圆脸	round face
鹅蛋脸	goose egg face
标准脸	standard face
长脸	long face
方脸	square face
倒鹅蛋脸	inverted oval face
长方形脸	rectangular face
梯形脸	trapezoidal face
倒梯形脸	inverted trapezoid face
菱形脸	diamond face
五角形脸	pentagonal face
梨形脸	pear shaped face
心形脸	heart shaped face
倒三角脸	inverted triangle face
不规则脸	irregular face

与情感表达有关的	
高兴的	happy
兴奋的	excited
惊喜的	surprised
愉快的	delighted
兴高采烈的	elated
欣喜若狂的	overjoyed
生气的	angry
害怕的	afraid
厌恶的	disgusted
狂喜的	ecstatic
欣喜的	joyful
悲伤的	sad
沮丧的	depressed
悲哀的	mournful
忧郁的	melancholic
心碎的	heartbroken
愤怒的	angry
狂怒的	furious
发怒的	irate
愤慨的	incensed
恐惧的	fearful
恐慌的	terrified
惊恐的	horrified
吓呆的	petrified
焦虑的	anxious
担心的	worried
惊讶的	amazed
可恶的	hateful
残忍的	brutal
安宁的	peaceful
安静的	quiet
神秘的	mysterious
舒适的	cozy
冒险的	adventurous
哀伤的	distressed
不安的	uncomfortable
平静的	calm
满足的	satisfied
尴尬的	embarrassed
感激的	grateful
惋惜的	regretful
紧张的	tense
孤独的	lonely
美丽的	beautiful

与情感表达有关的	
英俊的	handsome
可爱的	cute
史诗般的	epic
纠缠的	tangly
交织的	intertwined
巨大的	gigantic
湿润的	moist
缥缈的	misty
惊吓的	scary
友善的	friendly
聪明的	clever
开朗的	optimistic
谦虚的	humble
善良的	kind
成熟的	mature
乐观的	optimistic
自信的	confident
热情的	passionate
冷静的	calm
勇敢的	brave
活泼的	lively
独立的	independent
诚实的	honest
宽容的	tolerant
深情的	affectionate
心灵手巧的	ingenious
忠诚的	loyal
细心的	observant
悲观的	pessimistic
悲痛的	sorrowful
幸福的	happy
焦急的	anxious
震惊的	shocked
同情的	sympathetic
厌恶的	disgusted
悲痛的	sorrowful
陶醉的	intoxicated
贪婪的	greedy

与动作有关的	
走路	walk
跑步	running
跳跃	jump
蹦跳	crowhop
慢走	saunter
跳跃	leap
阔步行走	stride
蹒跚	stagger
跳高	high jump
跳远	long jump
坐	sit
站立	stand up
躺下	lie down
坐在地上	sit on the floor
站得笔直	stand up straight
直立行走	walk upright
坐直	sit up straight
屈膝坐下	sit down on knees
坐下来休息	sit down and rest
挥手	wave
拍手	clap
握手	shakehand
做手势	gesture
举手	raise one's hand
打招呼	greet
伸出手来	reach out
指着某物	point at something
抱紧	hold tightly
吃东西	eat
喝水	drink water
睡觉	sleep
唱	sing
跳舞	dance
游泳	swim
打篮球	play basketball
打乒乓球	play table tennis
打羽毛球	play badminton
开汽车	drive a car
骑自行车	ride a bike

与场景有关的	
充满生命力的雨林	life-affirming rainforest
人迹罕至的高山	off-the-beaten-track mountain
险峻的峭壁	precipitous cliff
神秘的森林	mysterious forest
宏伟的山脉	magnificent mountain
海底世界	underwater world
火山喷发	volcanic eruption
极地冰川	polar glacier
异星生物群落	alien biome
未知星球的奇异景观	strange landscape of unknown planet
沙漠中的神秘遗迹	mysterious ruins in desert
丛林中的危险生物	dangerous creatures in the jungle
冰冷的卫星表面	cold satellite surface
绝壁悬崖	sheer cliff
暴风雪中的零下世界	sub-zero world in blizzard
人工环境恒温室	artificial environment constant greenhouse
漆黑的夜晚	dark night
浓雾笼罩的城市	foggy citie
污浊的空气	dirty air
光滑的冰面	smooth ice
降雪的城市	snowy city
热浪袭来的空气	heat waves hitting the air
充满毒气的沙暴	toxic sandstorm
被污染的湖泊	polluted lake
狂风暴雨	violent storm
黑夜降临	night falls
星空闪耀	starry sky is shining
大雾弥漫的城市	foggy city
日落	sunset
彩虹出现	rainbows appearance
电闪雷鸣	lightning and thunder storms
风暴来袭	storm hits
海浪拍岸	waves lapping at the shore
热浪袭来	heat waves
沙尘暴	dust storms
寒冬腊月	cold winter
秋高气爽	clean and crisp autumn
炎炎夏日	hot summer
潮湿的雾气	humid fog
落叶满地	falling leaves
充满阳光	full of sunshine
杂草丛生	weeds spring up
梦境	dreamland
黑洞	black hole

与场景有关的	
银河	galaxy
星云	nebula
外行星	outer planet
超市	supermarket
医院	hospital
学校	school
公园	park
公司	company
马路	road
山村	village
田野	field
花园	garden
公路	highway
篮球场	basketball court
办公室	office
家	home
商场	shopping mall
银行	bank
地铁站	subway station
公交车站	bus stop
停车场	parking lot
快餐店	snack bar
幼儿园	kindergarten
小学	primary school
游乐场	playground
篮球馆	basketball hall
网球场	tennis court
食堂	canteen
图书馆	library
博物馆	museum
美术馆	art gallery
科技馆	science museum
大学	university
迪士尼乐园	disneyland
旅游景点	tourist attraction
长城	the great wall
故宫	the forbidden city
天坛	temple of heaven
健身馆	gymnasium
游泳馆	swimming pool
城市	city
近未来都市	near-future city
街景	street view
农田	farmland

与场景有关的	
草原	steppe
果园	orchard
冰山	iceberg
雪山	snowy mountain
冰川	glacier
峡谷	canyon
沼泽地	marshland
雨林	rainforest
河口	estuary
海底	seabed
火山	volcano
高山	high mountain
沙滩	sandbeach
山脉	mountains
废墟	ruins
教室	classroom
卧室	bedroom
森林	forest
海滩	beach
沙漠	desert
河流	river
湖泊	lake
瀑布	waterfall
珊瑚礁	coral reef
洞穴	cave
水下洞穴	underwater cave
城堡中的地牢	dungeon
天空之城	castle in the sky
奇幻森林	enchanted forest
城市公园	city park
深海	deep sea
巴比伦空中花园	hanging garden of babylon
珠穆朗玛峰	mount qomolangma
时空隧道	time tunnel
宇宙	universe
雨天	rainy day
时空裂缝	rift

与街道有关的

拥挤的街道	crowded street
灯火通明的商业街	iluminated shopping street
夜幕降临的街角	street corner at night
无人的马路	deserted road
覆盖着标语的隧道	tunnel covered with slogans
汽车穿梭的拥挤街道	crowded streets with cars weaving through
数不清的被霓虹灯点缀的商业街	countless commercial streets decorated with neon lights
迷雾笼罩的街角	street corners shrouded in mist
废弃的高速公路	deserted highway
墙壁和隧道上覆盖着五颜六色的涂鸦	walls and tunnels covered with colourful graffiti
高耸的桥梁连接了城市中的两个区域	towering bridge connects two areas of the city
废弃工业区的隧道	tunnels in abandoned industrial areas
密林中的荒道	deserted roads in dense forest
天然林中的山路	mountain roads in natural forest
带着有未知符号的神秘道路	mysterious road attached with unknown symbols
废弃的车站	abandoned station
黑暗的小巷	dark alley
升起的烟雾	rising smoke
破败的停车场	dilapidated car park
狭窄的小路	narrow lane
喧嚣的市场	noisy market
金融中心的大道	avenue of financial centre
工业区的道路	road in industrial area
地下车库	underground garage
陌生的城市	strange city
荒凉的大道	deserted avenue
漆黑的高速公路	dark highway
曲折的山路	winding mountain roads
隧道口	tunnel entrance
公路桥	highway bridge
水下隧道	underwater tunnel
飞天高速路	flying highway
虚拟道路	virtual road
热闹的街道上人来人往	people come and go on the busy street
街两旁种满了树木	streets are lined with trees
街头巷尾摆满了各种小吃摊位	streets and alleys are full of various food stalls
街道很宽阔，车辆行驶很顺畅	streets are wide and the traffic is smooth
街灯下行人正忙	pedestrians are busy under the street lights
在街道上可以感受到季节的变化	the change of seasons can be felt on the streets
街道被美丽的花坛装点着	streets are decorated with beautiful flower beds
街头有很多店铺和店面	many shops and storefronts on the street
晚上的街道格外迷人	streets at night are especially charming
在街上漫步是一种享受	roaming in the street for pleasure
路面泛着湿润的光泽	road has a wet sheen

与建筑有关的	
高耸的摩天大楼	towering skyscraper
废弃的工厂	abandoned factory
破败的公寓	dilapidated apartment
荒废的仓库	deserted warehouse
未来主义建筑	futuristic buildings
高达 100 层的摩天大楼	skyscrapers up to 100 storeys high
废弃的化工厂	abandoned chemical plants
废弃的地铁站	abandoned underground station
钢铁巨兽的未来主义建筑	futuristic building of steel giant
独特的废弃矿井	unique abandoned mine
城市中心的高桥	high bridge in the centre of city
破败的仓库区	dilapidated warehouse area
未来主义立交桥	futuristic overpasses
巨型城市塔楼	giant urban tower
高层公寓	high-rise apartment
混凝土丛林	concrete jungle
废弃的办公大楼	abandoned office block
高科技实验室	high-tech laboratory
宏伟的桥梁	magnificent bridge
巨大的地下设施	huge underground facility
发电厂	power station
防空洞	air-raid shelter
高速公路桥	motorway bridge
废弃的工业区	abandoned industrial area
未来主义的宇宙站	futuristic space station
高科技商业中心	high-tech commercial centre
科技领域的大型研发基地	large r&d base in the field of science and technology
高科技军事基地	high-tech military base
科技学院	science and technology college
炼油厂	oil refinery
气象站	weather station
军火库	arsenal
研究所	research institute
雄伟壮观的建筑	majestic building
独特的建筑设计	unique architectural design
精致的建筑细节	exquisite architectural detail
古老的建筑风格	old architectural style
现代风格的建筑	modern style building
富有历史感的建筑	historic building
美丽的建筑外观	beautiful building appearance
宏伟的建筑结构	magnificent architectural structure
壮观的建筑立面	spectacular building elevation
充满艺术感的建筑	building full of artistic sense

与交通有关的	
自行车	bike
摩托车	motorcycle
公共汽车	bus
三轮车	tricycle
电动车	electrocar
出租车	taxi
邮轮	cruise ship
帆船	sailboat
皮艇	kayak
潜艇	submarine
气垫船	hovercraft
飞机	airplane
宇宙飞船	spacecraft
火车	train
轻轨	light rail
地铁	subway
磁力悬浮火车	maglev
缆车	cable car
飞行汽车	flying car
巨型货车	giant lorry
电动自行车	electric bicycle
无人驾驶的出租车	driverless taxi
垃圾处理车	rubbish disposal vehicles
带有喷气式推进器的飞行汽车	flying cars with jet propulsion
覆盖着金属装甲的巨型货车	giant lorries covered with metal armour
速度极快的电动自行车	fast electric bicycle
自主行驶的公共汽车	autonomous bus
超级跑车	supercar
智能汽车	smart car
高速摩托车	high-speed motorbike
高科技悬浮车	high-tech hovercar
疾速地铁	speedy subway
空中飞行器	aerial vehicle
遥控卡车	remote-controlled truck
高科技飞艇	high-tech dirigible
科技制造的自行车	bicycle made of technology
宇宙战舰	space battleship
交通监控无人机	traffic surveillance drone
超级磁悬浮列车	super maglev train
尖端纯电动车	cutting-edge electric vehicle
超级飞行器	super flying machine
高科技水上交通工具	high-tech water vehicle

与科技有关的	
合成人类	synthetic human
机器人助手	robotic assistant
仿生生物	bionic being
智能宠物	intelligent pet
克隆人类	cloned human
合成人类的身体优势	physical advantages of synthetic humans
机器人助手的灵活性	flexibility of robotic assistants
仿生生物的自我进化	self-evolution of bionic beings
智能宠物的陪伴感	companionship of intelligent pets
克隆人类的完美基因	perfect genes of cloned humans
混合生物的多样性	diversity of hybrid beings
强化人类的身体能力	physical capabilities of enhanced humans
多重人格的思维方式	multiple personalities of thinking patterns
未来生命体的神秘力量	mysterious powers of future life forms
神秘生物的未知威胁	unknown threat from mysterious creatures
仿生机器人的情感	emotions of bionic robots
智能宠物的互动性	interactivity of intelligent pets
克隆人的道德争议	moral controversies of human cloning
合成人类的自我意识	self-awareness of synthetic humans
机器人助手的便利性	convenience of robotic assistants
仿生生物的生物力学特性	biomechanical properties of bionic beings
智能机器人的学习能力	learning capabilities of intelligent robots
虚拟人的真实感	realism of virtual humans
未来主义人类的演化	evolution of futuristic humans
机械生命体的意识认知	consciousness perception of mechanical life forms
人工智能的道德问题	moral issues of artificial intelligence
仿生人类的生理构造	physiological constitution of bionic humans
机器人研究的进展	advances in robotics research
人造生命的探索	the quest for artificial life
人工智能的学习方法	learning methods in artificial intelligence
智能机器人的自我保护能力	self-preservation capabilities of Intelligent robots
人类基因编辑的伦理问题	ethical issues in human gene editing
合成生命体的可控性	controllability of synthetic life forms
机器人工厂	robotic factory
虚拟现实技术	virtual reality
智能家居系统	intelligent home system
人工智能的无人机	artificial intelligence drone
具有自我学习能力的机器人工厂	self-learning robotic factory
能够沉浸式体验的虚拟现实技术	virtual reality technology for immersive experience
全方位智能家居系统	all-encompassing intelligent home system
具有突破性能力的电子设备	electronic devices with breakthrough capability
拥有卓越操作体验的便携式电脑	portable computers with superior operating experience
人类基因组改造技术	human genome modification technology
创造生命的合成技术	life-creating synthetic technology

与科技有关的	
能够实现时间穿越的科技	technologies that enable time travel
未来主义制造工艺	futuristic manufacturing process
无人化生产流水线	unmanned production line
可重复使用火箭	reusable rocket
人工智能大脑	artificial intelligence brain
虚拟现实设备	virtual reality device
多用途机械臂	multi-purpose robotic arm
深度学习算法	deep learning algorithm
无人驾驶技术	driverless technology
可持续能源科技	sustainable energy technology
先进的材料科学	advanced material science
量子计算机技术	quantum computer technology
黑科技武器	black technology weapon
生物工程技术	bio-engineering technology
生命科学研究	life science research
宇宙探索技术	space exploration technology
环境保护科技	environmental protection technology
高效节能技术	energy efficient technology
智能穿戴设备	intelligent wearable device
电子设备	electronic device
镭射眼镜	laser glasses
反光背心	reflective vest
机械手臂	robotic arm
智能化的战斗服	intelligent combat suit
改装后的时尚机甲	modified fashionable mech
磨砂黑夹克	frosted black jacket
黑色超短裙	black ultra short skirt
钢板战甲	plate armour
未来感发光鞋	futuristic glowing shoes
脚踝机械零件	ankle mechanical parts
半透明头盔	translucent helmet
智能手表	smart watch
光学护目镜	optical goggles
仿生战甲	bionic armour
智能眼镜	intelligent glasses
机械背包	mechanical backpack
防弹夹克	bulletproof jacket
仿生人造手臂	bionic artificial arm
高科技项链	high-tech necklace
光学瞄准镜	optical sight
星际飞船	starship
行星探测器	planetary probe
星系之间的太空旅行	intergalactic space travel
外星生命体	alien life form

与科技有关的

星际飞船的科技装备	technological equipment for starship
行星探测器的探测工具	probing tool for planetary probe
星系之间的太空旅行的未知冒险	unknown adventure of intergalactic space travel
外星生命体的威胁与猎捕	threat and hunting of alien life forms
星际贸易的繁荣与危机	prosperity and crisis of interstellar trade
星际战争的血腥与残酷	bloodshed and cruelty of interstellar war
太空探险的创新与进步	innovation and progress of space exploration
黑洞的奇妙力量与恐怖危险	wonderful power and terrible danger of black holes
未知星球的神秘环境与生命	mysterious environment and life on unknown planet
行星环境的探索	exploration of planetary environment
太空旅行的冒险	adventure in space travel
星系间的交通系统	intergalactic transportation system
奇怪的外星怪物	strange alien monsters
未知星球的探测	exploration of unknown planet
人类在宇宙中的生存	human survival in the universe
太阳系的演化过程	evolution of the solar system
太空站的生命保障系统	life support system on space station
地外文明的探索	exploration of extraterrestrial civilisation
恒星飞船的设计	design of stellar spaceship
黑洞的奥秘	mystery of black holes
宇宙中的暗物质	dark matter in the universe
星际探险队的挑战	the challenges of interstellar expeditions
星际战争的爆发	outbreak of interstellar war
科技设备在太空中的应用	use of technological devices in space
星际旅行的限制	limit of interstellar travel
未来的星际战略	interstellar strategy in the future
高科技能量场	high-tech energy field
激光束	laser beams
电磁脉冲	electromagnetic pulses
核反应堆	nuclear reactor
超级电池	super battery
电磁脉冲的瘫痪力	paralysing power of electromagnetic pulses
核反应堆的能量源	energy source of nuclear reactor
超级电池的持久力	staying power of super battery
量子力学的变幻力量	shifting power of quantum mechanics
黑洞的引力	gravitational power of black holes
核聚变的能量释放力量	energy-releasing power of nuclear fusion
电磁风暴的毁灭力量	destructive power of electromagnetic storms
高速磁力驱动的力量	power of high-speed magnetic drive
能量场的波动	fluctuations in energy field
电子设备的节能模式	energy-saving mode of electronic device
能量场的屏蔽效应	shielding effect of energy field
高科技武器的杀伤力	lethality of high-tech weapon
机械臂的承重能力	weight-bearing capacity of mechanical arm

与科技有关的

科技设备的耗电量	power consumption of technological equipment
核反应堆的输出能力	output of nuclear reactor
电磁脉冲的破坏力	destructive power of electromagnetic pulses
飞船发动机的推力	thrust of spaceship engine
高科技燃料的能量密度	energy density of high-tech fuels
太阳能发电的效率	efficiency of solar power generation
高科技设备的稳定性能	stability of high-tech equipment
能量转换的效率	efficiency of energy conversion
核聚变的热释放量	heat release of nuclear fusion
高科技设备的传输效率	transmission efficiency of high-tech equipment
能量流动的稳定性	stability of energy flow
穿越时空的旋涡	through the vortexe of time and space
虚拟现实空间	virtual reality space
神秘的传送门	mysterious portals
虚拟现实空间的幻境	illusions in virtual reality space
错乱的现实世界	the dislocated real world
未知力量的神秘现象	mysterious phenomena of unknown force
远古遗迹的探索之旅	voyage of exploration through ancient ruin
黑暗维度的恐怖体验	horrific experiences in dark dimension
未知星球的探险旅程	expedition to unknown planet
时空隧道的扭曲	warping of space-time tunnel
虚拟世界的幻觉	illusions of virtual world
科技外星人的出现	emergence of technological alien
未知星球的探索	exploration of unknown planet
仿生生命的逆转行为	reversal of the behaviour of bionic beings
高科技设备的错误反应	erroneous reaction of high-tech device
科技危机的爆发	outbreak of technological crisis
时空旅行的冒险	adventure of time travel
人类进化的奇妙之路	fantastic path of human evolution
科技研究的不可预知性	unpredictability of technological research
宇宙空间的磁场扰动	magnetic disturbance of cosmic space
虚拟人格的存在感	presence of the virtual personality
高科技设备的暴走行为	stormy behaviour of high-tech device
时空隧道的变异演化	mutant evolution of the space-time tunnel
未知的宇宙	unknown universe
超光速	faster-than-light
外星文明	alien civilization
未来世界	future world
人工智能	artificial intelligence
虚拟现实	virtual reality
平行宇宙	parallel universe
时间旅行	time travel
外星生物	aliens
量子世界	quantum world

与生活有关的	
电视机	television
电脑	computer
冰箱	refrigerator
洗衣机	washing machine
台灯	table lamp
电风扇	electric fan
灯具	lamp
沙发	sofa
桌子	table
椅子	chair
茶几	coffee table
化妆品	cosmetic
口红	lipstick
瓶子	bottle
男鞋	men's shoes
女鞋	women's shoes
旅游鞋	travel shoes
帆布鞋	canvas shoes
皮鞋	leather shoes
高跟鞋	high heels
婴儿车	baby carriage
汽车	car
电磁炉	induction cooker
牙刷	toothbrush
洗发水	shampoo
牙膏	toothpaste
沐浴露	body wash
护肤品	skin care product
床	bed
床垫	mattress
地毯	carpet
装饰画	decorative painting
水杯	water cup
炒菜锅	wok
笔	pen
笔记本	notebook
文件夹	folder
打印机	printer
纸张	paper
订书机	stapler
音响	audio
玩具	toy

与行业有关的	
零售业	retail industry
制造业	manufacturing industry
服务业	service industry
医疗保健业	healthcare industry
教育业	education industry
媒体和娱乐业	media and entertainment industry
餐饮业	catering industry
旅游业	tourism
建筑业	construction industry
物流和运输业	logistics and transportation industry
金融业	financial industry
农业	agriculture
科技产业	technology industry
电力和能源业	power and energy industry
矿业和采石业	mining and quarrying
建筑和房地产业	construction and real estate
政府和非营利组织	government and non-profit organization
汽车和交通工具制造业	automobile and vehicle manufacturing industry
航空和航天业	aviation and space industry
化工和制药业	chemical and pharmaceutical industry
环保行业	environmental industry
船舶和海洋工程行业	shipbuilding and offshore engineering industry
印刷和出版业	printing and publishing
建材和家居用品行业	building material and household goods industry
纺织和服装行业	textile and apparel industry
化妆品和个人护理用品行业	cosmetic and personal care industry
人力资源和招聘行业	human resource and recruitment industry
咨询和管理服务行业	consulting and management service industry
法律和法律服务行业	law and legal service industry
高等科技与生命科学行业	advanced technology and life science industry
游戏和软件开发行业	game and software development industry
市场营销和广告行业	marketing and advertising industry
社交媒体和网络科技行业	social media and internet technology industry
体育和运动行业	sports and exercise industry
家庭保健和老年照护行业	home health and aged care industry
教育科技和在线学习行业	education technology and online learning industry
汽车和交通服务行业	automotive and transportation services industry
包装和物流行业	packaging and logistics industry
娱乐设施和主题公园行业	amusement settings and theme park industry
建筑工程和设计行业	construction engineering and design industry
舞台艺术和表演行业	stagecraft and performance
艺术和文化产业	arts and cultural industry
消费品和零销售服务行业	consumer goods and retail service industry
机械和工业自动化行业	machinery and industrial automation industry

与行业有关的

通信和网络服务行业	communication and network services industry
原材料和采购行业	raw material and procurement industry
包装和印刷设备制造业	packaging and printing equipment manufacturing industry
健康和健康管理行业	health and wellness management industry
酒店和旅游服务行业	hotel and tourism service industry
线上和移动支持支付行业	online and mobile support for the payment industry
汽车配件和维修服务行业	auto part and repair service industry
塑料和化学制品行业	plastic and chemical industry
食品和饮料制造业	food and beverage manufacturing industry
宠物和动物服务行业	pet and animal service industry
玩具和儿童用品行业	toy and children's product industry
土木工程和公共基础设施建设行业	civil engineering and public infrastructure construction industry
珠宝和奢侈品行业	jewelry and luxury industry
家居装饰和工艺品行业	home decoration and craft industry
人力资源管理和培训行业	human resource management and training industry
人工智能和机器学习行业	artificial intelligence and machine learning industry
医疗器械和设备行业	medical device and equipment industry
资产管理和投资服务行业	asset management and investment service industry
物联网和智能家居行业	internet of things and smart home industry
保险和风险管理行业	insurance and risk management industry
石油和天然气行业	oil and gas industry
化工和材料科学行业	chemical and material science industry
航空货运和快递服务行业	air cargo and courier service industry
职业培训和教育行业	vocational training and education industry
新闻和出版行业	news and publishing industry
健康科技和数字医疗行业	health tech and digital health industry
云计算和数据中心业务	cloud computing and data center business
社区和社交服务行业	community and social service industry
生命科学和生物技术行业	life science and biotechnology industry
音乐和音乐制作业	music and music production industry
农业科技和粮食生产行业	agritech and food production industry
人类行为和心理学行业	human behavior and psychology industry
玩具和娱乐产品制造业	toy and entertainment product manufacturing industry
农业机械和工具制造业	agricultural machinery and tool manufacturing industry
能源和可再生能源行业	energy and renewable energy industry
物流和供应链管理行业	logistics and supply chain management industry
物联网和智能制造行业	internet of things and smart manufacturing industry
电信和网络设备行业	telecommunication and network equipment industry
社会媒体和内容制作行业	social media and content production industry
儿童教育和在线学习行业	children's education and online learning industry
家居清洁和维护服务行业	household cleaning and maintenance services industry
车联网和自动驾驶行业	internet of vehicle and autonomous driving industry

与食品有关的	
复合饮料	compound beverage
机械制作的食物	food made by machinery
烤肉串	kebab
营养均衡的合成食品	nutritionally balanced synthetic food
口感丰富的复合饮料	rich tasting compound beverage
味道鲜美的机械制作食品	tasty food made by machinery
烤肉串上散发出的浓郁香气	rich aroma from the kebab
高科技酿造的啤酒	high-tech brewed beer
特制的能提供精神力量的饮料	special drink designed to provide spiritual strength
科技制造的果蔬汁	fruit and vegetable juice made by science and technology
加入了新的调味配方的速食品	quick meal with a new seasoning
生物制造的肉类	bio-manufactured meat
食品打印机	food printer
合成食品	synthetic food
全营养液体食品	whole nutritious liquid food
高科技咖啡机	high-tech coffee machine
人造肉汉堡	artificial meat burger
仿生制作的水果	biomimetic fruit
实验室中的蔬菜	vegetable in laboratory
高科技饮料机	high-tech drink machine
生化能量饮料	biochemical energy drink
智能厨房设备	intelligent kitchen equipment
天然合成的饮料	natural synthetic drink
科技餐具	technological tableware
复合食品添加剂	compound food additives
生物基质食品	bio-based food
未来主义饮食理念	futuristic eating concept
仿生人造膳食	bionic artificial meal
高科技烹饪技术	high-tech cooking technology
令人回味无穷的	evocative
细腻顺滑入口即化	delicate and smooth, melt in the mouth
让人垂涎欲滴	mouth watering
香气扑鼻	fragrant
柔软细腻口感绝佳	soft and delicate taste excellent
酸甜可口	sweet and sour
口感独特	unique taste
色泽鲜艳口感丰盛	bright color and rich taste
清香扑鼻让人心情愉悦	fragrance makes people feel happy
口感鲜美香气扑鼻	delicious and fragrant
每口都是满满的幸福感	every bite is full of happiness

与布料有关的	
棉	cotton
亚麻布	linen
黄麻	jute
纸莎草纤维	papyrus fabric
羊毛	wool
山羊绒	cashmere
丝绸	silk
马海毛	mohair
羊驼毛	alpaca
纱	yarn
碳纤维	carbon fibers
金属纱	metallic yarn
尼龙	nylon
聚酯纤维	polyester
丙烯酸纤维	acrylic
人造丝	rayon
弹力纤维	elastane fiber
莫代尔	modal
薄纱质感	gauze texture
莱卡	lycra
人造革	artificial leather
纺织布	textile
涤棉	polyester cotton
雪纺	chiffon
牛仔布	denim
仿真丝	imitated silk fabric
皮革布	leather cloth
塑料布	plastic sheeting
丝绒	velvet
洗衣	laundry
美兰	meilan
针织布	knitted fabric
荷叶边	ruffles
毛绸	woolen silk
珠片布	sequin cloth
蓝丝绸	blue silk

与木材有关的	
木头	wood
松树	pine
冷杉	fir
云杉	spruce
雪松	cedar
侧柏	arborvitae
落叶松	larch
落羽杉	swamp cypress
花旗松	douglas fir
柏树	cypress
刺柏	juniper
红杉	redwood
北美白蜡木	american white ash
北美红橡木	american red oak
胡桃木	walnut
黄杨木 / 北美鹅掌楸	yellow poplar wood
柚木	teak
桃花心木	mahogany
榉木	beech
白桦树	white birch
橡木	oak
橡胶木	rubber wood
黑檀木	ebony
黄檀木	rosewood
紫檀木	padauk
定向纤维板	oriented strand board
胶合板	plywood
木芯板	blockboard
薄片木皮	veneer
人造板	wood-based panel
桦木	birch
松木	pine wood
枫木	maple
樟木	camphor wood
黑胡桃木	black walnut

与矿物有关的	
金	gold
银	silver
铜	copper
铁	iron
黄铜	brass
青铜	bronze
锡	tin
铂	platinum
铋	bismuth
铝合金	aluminum alloy
石灰岩	limestone
页岩	shale
砂岩	sandstone
板岩	slate
燧石	chert
熔岩	lava
火山岩	volcanic rock
安山岩	andesite
花岗岩	granite
玄武岩	basalt
黑耀石	obsidian
大理石	marble
石英岩	quartzite
蛇纹岩	serpentinite
水磨石	terrazzo
石灰华	travertine
混凝土	concrete
瓷砖	tile
鹅卵石	cobblestone
紫水晶	amethyst
钻石	diamond
日长石	sunstone
红宝石	ruby
蓝宝石	sapphire
天青石	celestite
青金石	lasurite
月长石	moonstone
紫水晶	amethyst
黄水晶	citrine
水晶	crystal
石英	quartz
石榴石	garnet
金绿玉	chrysoberyl
电气石	tourmaline

与矿物有关的	
绿松石	turquoise
祖母绿	emerald
海蓝宝石	aquamarine
橄榄石	olivine
玛瑙	agate
蛋白石	opal
绿柱石	beryl
坦桑石	tanzanite
蓝铜矿	azurite
孔雀石	malachite
翡翠	jadeite / jade
煤玉	jet
硫	sulfur
莫桑石	moissanite
萤石	fluorite
摩根石	morganite
粉红尖晶石	pink spinel
煤	coal

十二生肖	
鼠	rat
牛	ox
虎	tiger
兔	hare
龙	dragon
蛇	snake
马	horse
羊	sheep
猴	monkey
鸡	cock
狗	dog
猪	boar

与材质有关的	
树脂	resin
黏土质感	clay texture
菌丝	mycelium
玻璃	glass
棉花	cotton
亚麻	linen
蕾丝	lace
天鹅绒	velvet
砂岩	sandstone
薄纸	tissue paper
瓷器	porcelain
青瓷	celadon
珐琅	enamel
金属漆质感	metallic paint texture
纹理质感	texture
皮毛质感	fur texture
雕刻质感	carved texture
金属质感	metallic texture
石质	stone texture
珠光质感	pearl luster texture
玻璃质感	glass texture
皮革质感	leather texture
棉质	cotton texture
水晶质感	crystal texture
塑料质感	plastic texture
沙质	sandiness
亚光质感	matte texture
珍珠质感	pearl texture
绸缎质感	silk texture
毛绒质感	fluffy texture
水波纹质感	water wave texture
石墨质感	graphite texture
竹子质感	bamboo texture
陶瓷质感	ceramic texture
青铜质感	bronze texture
砖石质感	brick texture
油漆质感	paint texture
草木质感	plant texture
石膏质感	gypsum texture
瓷釉质感	porcelain glaze texture
树皮质感	bark texture
肌肉质感	muscle texture
金箔质感	gold foil texture
羽毛质感	feather texture

与细节有关的	
光滑的	smooth
清晰的	clear
细腻的	delicate
精细的	sophisticated
平整的	flat
精密的	precise
线条流畅的	sleek
流线型的	streamlined
柔和的	soft
弯曲的	curved
线条优美的	graceful
粗糙的	rough
不规则的	irregular
粗大的	bulky
锋利的	sharp
多角的	multangular
充满动感的	dynamic
统一的	uniform
多样化的	varied
有机的	organic
抛光的	polished
细微的	subtle
纤细的	slender
细长的	lathy
线条细致的	intricate
流畅且柔和的	smooth and soft
优美且流畅的	graceful and fluid
曲线细腻的	delicately curved
弯曲但流畅的	curved but fluid
优美且精细的	beautiful and detailed
优美且清晰的	beautiful and clear
有纹理的	textured
有层次感的	layered
纹路自然的	natural pattern
明显而突出的	obvious and prominent
具有浮雕感的	embossed
具有雕刻感的	carved
细节的	detailed
光滑的金属表面	smooth metal surface
磨砂金属表面	frosted metal surface
银色金属外观	silver metal look
金色金属质感	golden metal texture

与太阳系有关的	
太阳	sun
月亮	moon
水星	mercury
金星	venus
地球	earth
火星	mars
木星	jupiter
土星	saturn
天王星	uranus
海王星	neptunu
冥王星	pluto

星座	
水瓶座	aquarius
双鱼座	pisces
白羊座	aries
金牛座	taurus
双子座	gemini
巨蟹座	cancer
狮子座	leo
处女座	virgo
天秤座	libra
天蝎座	scorpio
射手座	sagittarius
摩羯座	capricornus

与艺术有关的	
水墨画	ink painting
山水画	landscape painting
中国画	chinese painting
工笔国画	chinese elaborate-style painting
水彩儿童插画	watercolor children's illustration
纸上彩墨	color ink on paper
水粉画	gouache painting
油画	oil painting
水彩	watercolor
版画	printmaking
素描	sketch
铅笔画	pencil drawing
炭笔画	charcoal drawing
钢笔画	pcn drawing
宣传画	picture poster
粉笔画	chalk drawing
涂鸦	doodle
剪纸工艺	paper-cutting craft
剪纸	paper cuttings
中国剪纸	chinese paper-cut
衍纸艺术	paper quilling art
刺绣艺术	embroidery
水墨渲染	ink render
民族艺术	ethnic art
花卉图案	floral pattern
古典建筑	classical architecture
唐诗宋词	tang and song poetry
传统建筑	traditional architecture
书画	calligraphy and paintings
茶道	tea ceremony
传统戏曲	traditional opera
花鸟文化	flower and bird culture
建筑写生	architectural sketch
手稿	manuscript
陶艺	pottery
雕塑	sculpture
动漫	anime
漫画	manga
像素画	pixel painting
插画	illustration
肖像画	portraiture
矢量图案	vector pattern
矢量插画	vector illustration
浮世绘	ukiyo-e

与艺术有关的	
二次元	acgn
卡通	cartoon
吉祥物	mascot
摄影	photography
数字艺术	digital art
装置艺术	installation art
建筑艺术	architectural art
印刷	printing
金属工艺	metalworking
纺织品设计	textile design
法国艺术	french art
海报风格	poster style
概念艺术	concept art
建筑设计	architectural design
民族艺术	ethnic art
国家地理	national geographic
雕刻艺术	carving art
点艺术	dot art
线艺术	line art
像素艺术	pixel art
皮克斯趋势	pixar trend
街景	streetscape
概念图	concept graph
字符画	ascii art
伏尼契手稿	voynich manuscript
室内设计	interior design
角色概念艺术	character concept art
线稿风格	line style
中国风	chinoiserie
绘画艺术	painting art
雕刻艺术	sculpture art
影视艺术	film and television art
摄影艺术	photography art
产品设计	product design
室内设计	interior design
视觉效果设计	visual effects design
服装设计	fashion design
书法	calligraphy
中国民间艺术	chinese folk art
陶瓷设计	ceramic design
木版年画	new year wood-block print
花鸟画	flower and bird painting
工笔画	claborate-style painting
波普艺术	pop art

与艺术有关的	
包豪斯	bauhaus
极简主义	minlmalist
新中式风格	new chinese style
写实风格	realism
光学艺术	optical art
超写实主义	hyperrealism
超现实主义	surrealism
巴洛克	baroque
印象派	impressionism
新艺术	art nouveau
洛可可	rococo
文艺复兴	renaissance
野兽派	fauvism
立体派	cubism
抽象表现主义	abstract art
点彩派	pointillism
维多利亚时代	victorian
未来主义	futuristic
后印象主义	post-impressionism
3D 风格	3d style
现代风格	modern style
乡村风格	country style
迪士尼风	disney-style
哥特式阴郁现实主义	gothic gloomy realism
野兽派	fauvism
极简主义	minimalist
工业风格	industrial style
美式乡村风格	american country style
粗犷主义	brutalism
数字雕刻风格	digital carving style
潜意识风格	subconscious style
黑白风格	black and white style
新现实主义	neo-realism
建构主义	constructivist
创世纪风格	genesis style
日本海报风格	japanese poster style
桁缝艺术	quilted art
原画风格	original painting style
水墨风格	ink style
日本漫画风格	japanese manga style
中国画风格	chinese traditional painting style
扁平风格	flat style
矢量风格	vector style
工艺美术运动	arts and crafts movement

与艺术有关的	
瑞士国际主义	swiss internationalism
纽约平面设计派	new york graphic design school
波兰海报设计	poland poster design
美国图钉派	american pushpin school
欧洲视觉诗人派	europe visual poets school
怪诞海报风格	quirky poster style
孟菲斯派	memphis faction
后现代主义	postmodernism
古典中国风	classical chinese style
国潮风格	chinese trendy style
2.5D 风格	2.5d style
故障风格	fault style
剪纸风格	pappersklippt stil
镭射风格	laserstil
酸性风格	syrlig stil
蒸汽波风格	steamwave style
赛博朋克	cyberpunk
纳米朋克	nanopunk
生物朋克	biopunk
黑暗赛博	cybernoir
蒸汽朋克	steampunk
时钟朋克	clockpunk
柴油朋克	dieselpunk
德科朋克	decopunk
原子朋克	atompunk
钢铁朋克	steelpunk
岛屿朋克	islandpunk
海洋朋克	oceanpunk
洛可可朋克	rococopunk
石头朋克	stonepunk
神话朋克	mythpunk
现代朋克	nowpunk
后赛博朋克	postcyberpunk
太阳朋克	solarpunk
月球朋克	lunarpunk
精灵朋克	elfpunk
霓虹朋克	neonpunk
赛博格	cyborg
废土朋克	wasteland punk
A 站趋势风格	station a trend style
Q 版可爱风	cute style
ISO 印刷风	iso printing style

与传统元素有关的	
汉服	hanfu
旗袍	cheongsam
中国龙	chinese dragon
中国凤凰	chinese phoenix
麒麟	kylin
中国灯笼	chinese lanterns
功夫	kung fu
咏春拳	wing chun
武侠	wuxia
昆曲	kunqu opera
笛子	flute
玉	jade
景泰蓝	cloisonne
瓷器	porcelain
刺绣	enbroidered
园林	gardens
亭子	pavilion
寺庙	temple
紫禁城	forbidden city
颐和园	summer palace
牡丹	peony
梅花	plum
莲花	lotus
竹子	bamboo
云纹	cloud pattern
金丝绒	pleuche
木纹	wood grain
琉璃	glass
古筝	guzheng
鸳鸯	mandarin ducks
扇子	fan
龙凤呈祥	dragon and phoenix
塔楼	tower
长衫	gown
绸缎	silk
文房四宝	four treasures of the study
砚台	inkstone
荷花	lotus
红楼梦	dream of the red chamber
古琴	gugin
佛像	buddha statue
围棋	go
古代钱币	ancient coins
凤凰	phoenix

与传统元素有关的	
四季景观	four seasons landscape
唐装	tang suit
狮子	lion
九尾狐	nine-tailed fox
狻猊	suan ni
貔貅	pi xiu
龙龟	dragon turtle
饕餮	gluttony
五行	five elements
中医	traditional chinese medicine
传统节日	traditional festival
传统建筑	traditional building
饮食文化	food culture
龙纹	dragon pattern
翎毛	teathers
唐三彩	tang sancai
禅意	zen
兰花	orchid
菊花	chrysanthemum
象形文字	hieroglyphs
福字	fu word
对联	couplet
花卉纹	flower pattern
窗花图案	window pattern
太极图案	tai chi pattern
莲花图	lotus flower
宫殿	palace
明式家具	ming style furniture
中国结	chinese knot
京剧	peking opera
屏风	byobu
寿字	shou character
蝙蝠纹	bat pattern
瑞兽	rui beast
云头纹	cloud pattern
牡丹纹	peony pattern
中国的神明	chinese gods and immortals
中国神话	chinese mythology
哮天犬	dog in chinese mythology
狐狸精	vixen
石狮子	stone lion
天池水怪	lake tianchi monster
月兔	moon rabbit
山海经	mountains and seas sutra

与传统元素有关的	
神兽	mythical beasts
蚩尤	chi you
九头鸟	nine-headed bird
穷奇	qiongqi
朱雀	vermilion bird
妖怪	monster
儒家文化	confucian culture
道教	daoism
刺绣	embroidery
文物	heritage
莫高窟	mogao caves
宣纸	chinese art paper
三国演义	romance of the threekingdoms
茶道	sado
印章	seal
四合院	siheyuan
宋朝	song dynasty
年画	spring festival painting
太极拳	tai chi
唐朝	tang dynasty
寺院	temple
兵马俑	terracotta warriors
诗经	the book of songs
江湖	the jianghu world
西游记	the journey to the west
五行说	theory of five elements
传统国画	traditional chinese painting
八卦	trigram
战国	warring states
水浒传	water margin
阴、阳	yin, yang
禅宗	zen buddhism
中国脸谱	chinese opera mask
本草纲目	compendium of materia medica
火药	gunpowder
史记	historical records
农历	chinese calendar
孟子	mencius
冰糖葫芦	a stick of sugar-coated haws
八宝饭	eight-treasure rice pudding

概念分类	
专辑封面	album cover
解剖图	anatomical drawing
解剖学	anatomy
书籍封面	book cover
品牌标识	brand ldentity
名片	business card
日历设计	calendar design
角色设计	character design
多种姿势的角色设计	character design multiple poses
角色卡	character sheet
图表设计	chart design
颜色调色板	color palette
填色书页	coloring book page
连环画	comic strips
主视觉	master vision
作品集	portfolio
百科全书	encyclopedia
时尚想法模板	fashion idea templates
平面摄影	flat lay photography
传单	flyer
全身人物设计	full body character design
游戏资产	game assets
品牌识别	brand recognition
图标设计	icon design
图标集设计	icon set design
四方连续插图	square series
标志设计	logo design
杂志封面	magazine cover
菜单设计	menu design
App UI 设计	mobile app UI design
包装设计	packaging design
海报设计	poster design
宣传海报	propaganda poster
字体设计	font design
版面设计	layout
网格系统	grid system
色彩方案	color scheme
留白	white space
构图	composition
视觉层次	visual hierarchy
品牌设计	branding
模型图	diagrammatic figure
矢量图形	vector graphics
位图	raster graphics

概念分类	
出血	bleed
印刷四色	cmyk
网页	web page
首页	homepage
导航	navigation
横幅广告	banner
按钮	button
链接	link
菜单	menu
下拉菜单	dropdown menu
轮播图	carousel
响应式设计	responsive design
用户界面	user interface
用户体验	user experience
网站图标	website icon
线框图	wireframe
原型	prototype
汉堡菜单	hamburger menu
错误页面	error page
可用性测试	usability testing
信息架构	information architecture
色彩搭配	color scheme
背景设计	background design
界面布局	interface layout
导航设计	navigation design
视觉设计	visual design
原型设计	prototype design
设计规范	design standards
排版	typography
分辨率	resolution
像素密度	pixel density
对比度	contrast
对齐	alignment
比例	proportion
模板	template
视觉识别	visual identity
设计简报	design brief
无衬线字体	sans-serif
黄金时段光	golden hour light
赛博朋克光	cyberpunk light
柔软的光线	soft light
明亮的光线	bright light
聚光灯	spot light
环形灯	ring light

概念分类	
纹理灯	textured light
面板灯	panel light
频闪灯	strobe light
霓虹灯	neon light
冷光霓虹灯	neon cold lighting
荧光灯	fluorescent lighting
边缘灯	edge light
荧光灯	fluorescent lighting
气氛照明	atmospheric lighting
情绪照明	mood lighting
双性照明	bisexual lighting
伦勃朗照明	rembrandt lighting
分体照明	split lighting
背光照明	back lighting
全局照明	global lighting
正面照明	front lighting
背景照明	back lighting
室内照明	indoor lighting
室外照明	outdoor lighting
高动态范围照明	high dynamic range lighting
实时照明	real-time lighting
边缘照明	edge illuminated
干净的背景	clean background
点光源	point light
自发光材质	self-luminous material
黄昏射线	twilight ray
强烈光线	hard lighting
发光文本	glowing text
电弧火花	arc sparks
彩虹光环	rainbow halo
夜光效果	glow in the dark
激光束效果	laser beam effect
暗光效果	dim light effect
星空效果	starry sky effect
头发发光效果	hair glow
丁达尔光	tyndall light

光线词汇	
色光	color light
自然光	natural light
太阳光	sun light
暖光	warm light
冷光	cold light
强光	highlight
硬光	hard light
柔和光	soft illumination
双面性灯光	double-sided lighting
晨光	dawn
逆光	back light
侧光	raking light
顶光	top light
立体光	stereo luminescence
反光	reflection light
层次光	level light
轮廓光	rimming light
边缘光	edge light
球形灯光	spherical light
半球灯光	hemisphere light
区域灯光	area light
环境光	ambient light
阴影光	shadow light
微光	rays of shimmering light
映射光	mapping light
情调光	mood lighting
工作室灯光	studio lighting
戏剧性灯光	dramatic lighting
定向光	directional light
分光灯光	split lighting
霓虹灯冷光	neon light
体积光	volumetric lighting
电影光	cinematic light
影棚光	studio light
电影灯光	dramatic lighting
耀眼的光线	rays of shimmering light
情境灯光	mood lighting
暮光	twilight
好看的灯光	beautiful lighting

光线氛围	
梦幻雾气	dreamy haze
明暗分明	clear light and dark
黑暗氛围	dark atmosphere
鲜艳色彩	vibrant color
高对比度	high contrast
安静恬淡	serene calm
明亮高光	bright highlights
闪耀星空	twinkling stars
柔和烛光	soft candlelight
暧昧光晕	ambiguous halo
自然光	natural light
魔法森林	enchanted forest
仙气缭绕	ethereal mist
温暖光辉	warm glow
忧郁氛围	melancholy atmosphere
柔和月光	soft moonlight
微光	shimmer
浪漫烛光	romantic candlelight
闪耀的霓虹灯	shimmering neon lights
黑暗中的影子	shadow in the dark
照亮城市的月光	moonlight illuminating the city
强烈的阳光	strong sunlight
熠熠生辉的霓虹灯	glittering neon lights
黑暗中的神秘影子	mysterious shadow in the dark
折射光线下的变幻光影	changing light and shadow under refracted light
闪烁不定的烛光	flickering candlelight
星光下的美丽影像	beautiful images in starlight
柔和的阴影	soft shade
梦幻般的光影效果	dreamy light effects
烟雾中的迷离影像	misty images in smoke
未来主义的夜景	futuristic night scenes
红色的霓虹灯光	the red neon light
充满幻想的星空	fantasy starry skies
机器人的投影	projections of robots
未来的科技光束	beams of future technology
黑暗中的眼睛	eyes in the dark
闪耀的星星	shining stars
强烈的太阳光线	intense sun rays
电影中的未来世界光影	light and shadow in the future world in films
虚拟现实中的光影	light and shadow in virtual reality
高科技眼镜的反射光	reflected light from high-tech glasses
未来世界中的阴影与光影	shadow and light in the future world
机器人身上的光线投影	light projections on robots

与构图有关的	
对称构图	symmetrical the composition
非对称构图	asymmetrical composition
三分法构图	rule of thirds composition
黄金分割构图	golden ratio composition
对角线构图	diagonal composition
动态对称构图	dynamic symmetry composition
并列构图	juxtaposition composition
汇聚线条构图	converging lines composition
消失点构图	vanishing point composition
画框构图	framing composition
非线性构图	nonlinear composition
饱和构图	saturated composition
视角构图	point of view composition
剪影构图	silhouette composition
重复构图	repetition composition
焦点构图	focal point composition
对比构图	contrast composition
重叠构图	overlapping composition
S 形构图	s-shaped composition
居中构图	center the composition
水平构图	horizontal composition
垂直构图	vertical composition
三角形构图	triangle composition
长方形构图	rectangular composition
圆形构图	circular composition
辐射构图	radiation composition
中心式构图	central composition
渐次式构图	progressive composition
散点式构图	scatter composition
平铺式构图	tile composition
倾斜视图	tilt view
跃肩视图	over-the-shoulder shot
俯视图	top view
正视图	front view
曼荼罗构图	mandala composition
对称面孔	symmetrical face
孤立构图	isolation composition
径向构图	radial composition
分割互补构图	split complementary composition
布景构图	scenery composition
负空间构图	negative space composition
拼贴构图	collage composition
微距拍摄	macro shot
广阔的视野	wide vision

与构图有关的	
人像	portrait
大头照	headshot
对称的身体	symmetrical body
对称	symmetry
曼德博集合	mandelbrot set
头部特写	headshot
极端特写	extreme closeup
黄金分割	golden cut
不对称	asymmetry
引导线	guide line
构框	framing
孤立	isolation
深度	depth
对比	contrast
纹理	texture
倾斜位移	tilt displacement
微观	microscopic view
两点透视	two-point perspective
三点透视	three-point perspective
立面透视	elevation perspective
横截面图	a cross-section view
电影镜头	cinema lens
焦点对准	in focus
景深	depth of field
细节镜头	detail shot
面部拍摄	face shot
异形构图	irregular composition
空间构图	spatial composition
曲线构图	curve composition
三角法构图	trigonometric composition
五分法构图	rule of five composition
重心法构图	center of gravity composition
透视法构图	perspective composition
装饰性结构图	decorative structure diagram
镜头选择	lens selection
空间关系	spatial relationship

与视角有关的	
自由视角	free perspective
固定视角	fixed perspective
俯视视角	top-down perspective
侧面视角	side-scrolling perspective
超级侧角	super side angle
广角镜头	wide- angle view
短距离视角	close-up view
远距离视角	long-shot view
随意视角	arbitrary view
全景视角	panoramic view
逆光摄影风格	backlight style
低角度视角	low angle shot
俯拍视角	overhead view
常规视角	conventional perspective
鸟瞰视角	bird's-eye view
交错视角	alternate angle of view
运动视角	motion perspective
超广角	ultra wide angle
广角	wide angle
透视视角	perspective
底特律视角	detroit perspective
肩膀视角	over the shoulder
鱼眼视角	fisheye lens
微距视角	macro lens
反转视角	reverse angle
跟随视角	following camera
内视镜视角	endoscopic view
第一人称视角	first-person view
第三人称视角	third-person perspective
仰视	look up
顶视	top view
俯视	look down
前视、侧视、后视图	foresight, side view, rear view
底视图	bottom view
卫星视图	satelite view
产品视图	product view
极端特写视图	extreme closeup view
等距视图	isometric view
特写视图	closeup view
高角度视图	high angle view
特写视图	closeup view
高角度视图	high angle view
鸟瞰图	aerial view
近距离景	close-up view

与视角有关的	
中景	medium shot
远景	long shot
前景	foreground
中远景	medium long shot
过肩景	shoulder view
松散景	loose shot
宽景	wide view
群景	group scenery
两景、三景	two shot, three shot
风景照	landscape photography
背景虚化	bokeh
全身	full body
半身像	bust
正面	front
侧面	profile
大特写	detail shot
特写	close-up
中特写	medium close-up
极限特写	limiting close-up
极限近景	extreme close-range
模拟摄影机	simulated camera
胸部以上	chest shot
腰部以上	waist shot
膝盖以上	knee shot
头部以上	big close-up
脸部特写	face shot
外太空视角	outer space view
高空视角	high altitude view
高视角	high-angle view
低视角	low-angle view
侧视角	side angle
低角度视角	low-angle shot
高角度视角	high-angle shot
俯视角度	top-down shot
斜角度	oblique angle
鸟瞰角度	bird's-eye view
主观视角	subjective perspective
客观视角	objective perspective
第一视角	first perspective

与色彩有关的	
金银色调	gold and silver tone
马卡龙色系	macaron color palette
明亮色系	bright color palette
暖色系	warm color palette
白色和粉红色调	white and pink tone
黄黑色调	yellow and black tone
红黑色调	red and black tone
霓虹色调	neon shades
黑色背景为中心	black background centered
多色彩搭配	colourful color matching
多彩的色调	rich color palette
亮度	luminance
低纯度色调	the low-purity tone
高纯度色调	the high-purity tone
淡色调	light color
单色调	monotone
对比度	contrast
红色	red
白色	white
黑色	black
绿色	green
黄色	yellow
蓝色	blue
紫色	purple
灰色	gray
棕色	brown
褐色	tan
青色	cyan
橙色	orange
水晶蓝色系	crystal blue
发光	shine
星闪	star flash
荧光	fluorescence
圣光	holy light
反射透明彩虹色	reflections transparent iridescent colors
糖果色系	candy color
珊瑚色系	coral color
紫罗兰色系	violet
玫瑰金色系	rose gold
浅蓝色系	light blue
酒红色系	burgundy
薄荷绿	mint green
枫叶红色系	maple red
雪山蓝色系	snow mountain blue

与色彩有关的	
白色和绿色调	white and green tones
红色和黑色调	red and black tones
黄色和黑色调	yellow and black tones
金色和银色调	gold and silver tones
镭射糖果纸色	laser candy paper color
莫兰迪色系	morandi color
钛金属色系	titanium color
黑白色系	black and white
极简黑白色系	minimalist black and white
温暖棕色系	warm brown
柔和粉色系	soft pink
时尚灰色系	fashionable gray
亮丽橙色系	bright orange
象牙白色系	Ivory white
自然绿色系	natural green
奢华金色系	luxurious gold
稳重蓝色系	steady blue
经典红黑白色系	classic red black and white
珊瑚橙色系	coral orange
秋日棕色系	autumn brown
丹宁蓝色系	denim blue
柠檬黄色系	lemon yellow
极致灰	extreme gray
炫彩黄	bright yellow
淡粉色	light pink
原色系	primary color system
土色系	earth tone
鲜艳色系	bright colors
柔和系	softness
自然系	natural
海洋系	marine
大地色系	earth tones
冰激凌色系	ice cream color

与画面精度有关的	
高细节	high detail
高品质	hyper quality
高分辨率	high resolution
全高清	fhd
虚幻引擎	unreal engine
建筑渲染	architectural visualisation
室内渲染	interior rendering
真实感	realistic
8K 流畅	8k smooth
8K 分辨率	8k resolution
16K 分辨率	16k resolution
超高清	ultra high definition
超清晰	super clarity
高清	high definition
超高分辨率	ultra-high resolution
超高清 HDR	ultra hd hdr
超高清晰度	ultra-high clarity
超高清晰	ultra-high definition
超高清画面	ultra hd picture
4K 画质	4k
8K 画质	8k
高清	hd
高动态光照渲染图像	hdr
超高清画质	uhd
摄影图片	photography
详细的细节	detailed
真实生活	real life
35 毫米镜头	35mm lens
现实主义的	realistic
游戏质感	game texture
3D 渲染	3d rendering
电影感	cinematic
徕卡镜头	leica lens
场景设计	scene design
照相写实主义	photorealistic
超真实的	hyper realistic
注重细节	high on details
电影效果	cinematic effect
空间反射	space reflection
错综复杂的细节	intricated details
商业摄影	commercial photography
渐进式渲染	progressive rendering

与画面渲染有关的	
OC 渲染	octane rendering
v 射线	v-ray
UE5 效果	unreal engine 5 effect
虚拟引擎	virtual engine
超真实	hyperrealism
环境光遮蔽	ambient occlusion
物理渲染	physically based rendering
景深	depth of field
抗锯齿	antialiasing
体积渲染	volume rendering
光线追踪	ray tracing
光线投射	ray cast
格点光线追踪	grid-based ray tracing
重要性采样	importance sampling
Monte Carlo 渲染	monte carlo rendering
贴图映射	texture mapping
环境贴图	environment map
全局光照	global illumination
着色器	shader
亚像素采样	subpixel sampling
Arnold 渲染器	arnold renderer
V-Ray 渲染器	v-ray renderer
Redshift 渲染器	redshift renderer
C4D 渲染器	c4d renderer
Blender 渲染器	blender renderer
Mental Ray 渲染器	mental ray renderer
靛蓝渲染器	indigo renderer
FStorm 渲染器	fstorm renderer
Corona 渲染器	corona renderer
纹理映射	texture mapping
渲染引擎	rendering engine
颗粒效果	particle effect
深度缓冲	depth buffer
动态模糊	dynamic blur
屏幕空间环境光遮蔽	screen space ambient occlusion, ssao
光渗漏	light bleed
间接光照	indirect lighting

艺术家

/ 意大利 /

乔托迪·邦多内	Giotto di Bondone
西莫内·马丁尼	Simone Martini
弗拉·安哲里柯	Fra Angelico
帕多·乌切洛	Pado Ucoello
秦梯利·德·法布里亚诺	Gentile de Fabriano
马萨乔	Masaccio
弗拉·菲利普·利比	Fra Fiippol lippi
安德烈亚·德尔·卡斯塔尼奥	Andrea del Castagno
贝诺佐·哥佐利	Benozzo Gozzoli
皮耶罗·德拉·弗朗切斯卡	Piero della Francesca
乔凡尼·贝利尼	Giovanni Bellini
安德烈亚·曼坦那	Andrea Mantegna
卡洛·克里韦利	Carlo Crivelli
桑德罗，波提切利	Sandro Botticelli
佩鲁吉诺	Perugino
多梅尼科·基兰达约	Domenico Ghirlandajo
列奥纳多·达·芬奇	Leonardo da Vinci
菲比皮诺·利比	Filippino Lippi
米开朗基罗·博那罗蒂	Michelangelo Buonarroti
乔尔乔涅	Giorgione
洛伦佐·洛托	Lorenzo Lotto
帕尔马·伊尔·韦基奥	Palma il Vecchio
拉斐尔·桑西	Raffaello Santi
安东尼奥·柯勒乔	Antonio Correggio
提香·韦切利奥	Tiziano Vecellio
多索·多西	Dosso Dossi
帕米贾尼诺	Parmigianino
布隆齐诺	Bronzino
丁托莱托	Tintoretto
乔凡尼·巴蒂斯塔·莫罗尼	Giovanni Battista Moroni
保罗·委罗内塞	Paolo Veronese
安德·德尔·萨尔托	Andrea del Sarto
安尼巴莱·卡拉齐	Annibale Carracci
奥拉齐奥·真蒂莱斯基	Orazio Gentileschi
米开朗基罗·梅西里·达·卡拉瓦乔	Michelangelo Merisi da Caravaggio
圭多·雷尼	Guido Reni
阿尔泰米西亚·真蒂莱斯基	Artemisia Gentileschi
詹巴蒂斯塔·提埃波罗	Giambattista Tiepolo
蓬佩奥·巴托尼	Pompeo Batoni
乔凡尼·塞冈提尼	Giovanni Segantini
阿梅代奥·莫迪利亚尼	Amedeo Modigliani
雷纳托·古图索	Renato Guttuso
胡伯特·凡·艾克	Hubet Van Eyck

艺术家	
扬·凡·艾克	Jan van Eyck
佩特吕斯·克里司图斯	Petrus Christus
罗吉耶·凡·德·威登	Pogier Van der Wenyden
迪里克·布茨	Dieric Bouts
罗贝尔·康宾	Robert Campin
汉斯·梅姆林	Hans Memling
雨果·凡·德·古斯	Hugo Van der Goes
希罗尼穆师·博斯	Hierongmus Bosch
盖特根·托特·辛特·扬斯	Geertgen tot Sint Jans
格拉德·戴维	Gerard David
康坦·马西斯	Quentin Massys
扬·戈塞特	Jan Gossaert
彼得·勃鲁盖尔	Pieter Bruegel
/ 法国 /	
弗朗索瓦·克鲁埃	Francois Clouet
安托万·勒南	Antoine Le Nain
路易·勒南	Louis Le Nain
马蒂厄·勒南	Mathieu Le Nain
西蒙·武埃	Simon Vouet
乔治·德·拉·图尔	Georges de La Tour
尼古拉斯·普桑	Nicolas Poussin
克洛德·洛兰	Claude Loorrain
菲利普·德·尚帕涅	Philippe de Champaigne
夏尔·勒·布伦	Charles Le Brun
厄斯塔什·勒·叙厄尔	Eustache Le Sueur
塞巴斯蒂安·布尔东	Sebastien Bourdon
让·弗朗索瓦·德·特洛伊	Jean-François de Troy
亚森特·里戈	Hyacinthe Rigaud
让·安托万·华托	Jean-Antoine Watteau
让·马克·纳蒂埃	Jean-Marc Nattier
弗朗索瓦·勒穆瓦纳	Francois Lemoyne
路易斯·托克	Louis Tocque
让·巴蒂斯特·西梅翁·夏尔丹	Jean-Baptiste Simeon Chardin
弗朗索瓦·布歇	Francois Boucher
克洛德·约瑟夫·韦尔内	Claude-Joseph Vernet
让·巴蒂斯特·格瑞兹	Jean-Baptiste Greuze
让·奥诺雷·弗拉戈纳尔	Jean-Honore Fragonard
雅克·路易·达维特	Jacqnes Louis David
巴龙·让·巴蒂斯特·勒尼奥	Beron Jean-Baptiste Regnault
皮埃尔·保罗·普吕东	Pierre-Paul Prud'hon
弗朗索瓦·热拉尔	Francois Gerard
安托万·让·格罗	Antoine Jean-Gros
弗朗索瓦·爱德华·皮科特	Francois Edouard Picot
阿道夫·威廉·布格罗	Adolphe William Bouguereau

艺术家	
泰奥多尔·席里柯	Theodore Gericaule
保尔·德拉罗什	Paul Dalaroche
让·巴蒂斯特·卡米耶·柯罗	Jean-Baptiste Camille Cont
欧仁·德拉克洛瓦	Euaene Delacroix
让·奥古斯特·多明尼克·安格尔	Jean-Auguste Domingue Engres
泰奥多尔·夏塞里奥	Theodore Chasserian
让·莱昂·热罗姆	Jean-Leon Gerome
泰奥多尔·卢梭	Theodore Rousseau
奥诺雷·杜米埃	Aonre Daumier
让·弗朗索瓦·米勒	Jean-Francois Millet
居斯塔夫·库尔贝	Gustave Courbet
罗莎·博纳尔	Rosa Bonheur
皮埃尔·皮维·德·夏凡纳	Pierre Puvis de Chavannes
古斯塔夫·莫罗	Gustave Moreau
朱尔斯·布勒东	Jules Breton
卡米耶·毕沙罗	Camille Pissarro
爱德华·马奈	Edouard Manet
朱尔斯·巴斯蒂安·勒帕热	Jules Bastien Lepaze
朱尔斯·勒菲弗尔	Jules Lefebvre
埃德加·德加	Edgar Degas
雅姆·蒂索	James Tissot
亨利·方丹·拉图尔	Henri Fantin Latour
艾尔弗雷德·西斯莱	Alfred Sisley
保罗·塞尚	Paul Cézanne
奥斯卡-克劳德·莫奈	Oscar-Claude Monet
皮埃尔·奥古斯特·雷诺阿	Pierre Augnste Renoir
贝尔特·莫里索	Berthe Morisot
让·勒孔特·迪·努伊	Jean-Lecomte du Nouy
亨利·卢梭	Henri Rousseau
古斯塔夫·卡耶博特	Gustave Caillebotte
保罗·高更	Paul Gauguin
乔治·修拉	Georges Senrat
亨利·热尔韦	Henri Gervex
亨利·玛丽·雷蒙·德·图卢兹-劳特累克	Henri Marie Raymond de Toulouse-Lautrec
亨利·马蒂斯	Henri Matisse
阿尔贝·马尔凯	Albert Marquet
莫里斯·德·弗拉芒克	Maurice de Vlaminck
安德烈·德朗	Andre Derainc
费尔南德·莱热	Fernand Leger
乔治·布拉克	Georges Braque
马塞尔·杜桑	Marcel Duchamp
马克·夏加尔	Marc Chagall
古斯塔夫·多雷	Gustave Dore
雅克·卡洛	Jacques Callot

艺术家

/ 西班牙 /

胡塞佩·德·里贝拉	Jusepe de Ribera
埃尔·格列柯	El Greco
迭戈·罗德里格斯·德·席尔瓦·委拉斯开兹	Diego Rodriguez de Silva y Velazquez
巴托洛梅·埃斯特万·穆立罗	Bartolome Esteban Murillo
弗朗西斯科·德·戈雅·卢西恩特斯	Fraycisco de Goyay Lucientes
巴勃罗·毕加索	Pablo Picasso
约安·米罗	Joan Miro
萨尔瓦多·达利	Salvador Dali

/ 德国 /

阿尔布雷特·丢勒	Albrecht Durer
马蒂亚斯·格吕内瓦尔德	Matthias Grunewald
卢卡斯·克拉纳赫	Lucas Cranach
阿尔布雷特·阿尔特多费尔	Albrecht Altdorfer
汉斯·荷尔拜因	Hans Holbein
约翰·海因里希·威廉·蒂施拜因	Johann Heinrich Wilhelm Tischbein
菲利普·奥托·龙格	Philipp Otto Runge
卡斯帕·达维德·弗里德里希	Caspar David Friedrich
阿道夫·门采尔	Adolf Menzel
威廉·莱伯尔	Wilhelm Leibl
保罗·克利	Paul Klee
马克斯·贝克曼	Max Beckmann
奥托·迪克斯	Otto Dix
马克思·恩斯特	Max Ernst
乔治·格罗兹	Georeg Grosz
卢西安·弗洛伊德	Lucian Frend

/ 荷兰 /

弗兰斯·哈尔斯	Frans Hals
威廉·克拉斯·赫达	Willem Claesz. Heda
让·凡·霍延	Jan van Goyen
伦勃朗·哈尔曼松·凡·莱因	Rembrandt Harmenszoon van Rijn
亚伯拉罕·凡·贝耶伦	Abraham van Beyeren
雅各布·凡·雷斯达尔	Jacob van Ruisdael
扬·斯汀	Jan Steen
彼得·德·霍赫	Pieter de Hooch
约翰内斯·维米尔	Johannes Vermeer
梅因德尔特·霍贝玛	Meindert Hobbema
文森特·威廉·梵高	Vincent Willem van Gogh
彼埃·蒙德里安	Piet Mondrian

/ 俄罗斯 /

阿·加·维聂茨昂诺夫	Alexey Gavrilovich Venetsianov
吉普林斯基	Orest Kiprensky
卡尔·巴甫洛维奇·布留洛夫	K·P·Brullov
亚历山大·伊凡诺夫	Alexander Ivanov

艺术家	
帕威尔·费多托夫	Pavel Fedotov
伊凡·康斯坦丁诺维奇·艾瓦佐夫斯基	Ivan Konstantinovich Aivazovsky
A·K·萨夫拉索夫	A·K·Savrasov
伊凡·伊凡诺维奇·希施金	Ivan I. Shishkin
伊里亚·叶菲莫维奇·列宾	Ilya Efimovich Repin
伊萨克·伊里奇·列维坦	Isaak Iliich Levitan
米哈伊尔·弗鲁贝尔	Mikhail Vrubel
瓦西里·康定斯基	Wassily Kandinsky
/ 英国 /	
威廉·霍加斯	Willian Hogarth
乔舒亚·雷诺兹	Joshua Reynolds
乔治·斯塔布斯	George Stubbs
托马斯·庚斯博罗	Thomas Gainsborough
约瑟夫·莱特	Joseph Wright
约翰·佐法尼	Johan Zoffany
托马斯·劳伦斯	Thomas Lawrence
詹姆斯·巴里	James Barrie
约瑟夫·马洛德·威廉·透纳	Joseph Mallord William Turner
约翰·康斯特布尔	John Constable
理查德·达德	Richard Dadd
约翰·马丁	John Martin
福特·马多克斯·布朗	Ford Madox Brown
威廉·霍尔曼·亨特	William Holman Hunt
约翰·埃弗里特·米莱斯	John Everett Milliais
爱德华·伯恩·琼斯	Edward Burne-Jones
阿瑟·休斯	Arthur Hughes
阿尔玛·塔德玛	Alma Tedema
弗朗西斯·培根	Francis Bacon
/ 美国 /	
温斯洛·霍默	Winslow Homer
詹姆斯·阿博特·麦克尼尔·惠斯勒	James Abbott McNeill Whistler
玛丽·卡萨特	Mary Cassatt
约翰·辛格·萨金特	John Singer Sargent
爱德华·霍普	Edward Hopper
威廉·德·库宁	Willem de Kooning
杰克逊·波洛克	Jackson Pollock
安德鲁·纽厄尔·怀斯	Andrew Newell Wyeth
安迪·沃霍尔	Andy Warhol
汤姆·韦塞尔曼	Tom Wesselmann
/ 日本 /	
手冢治虫	Osamu Tezuka
鸟山明	Akira Toriyama
宫崎骏	Miyazaki Hayao
高桥留美子	Rumiko Takahashi

艺术家	
富坚义博	Togashi Yoshihiro
伊藤润二	Junji Ito
小原古邨	Ohara Koson
奈良美智	Yoshitomo Nara
尾田荣一郎	Eiichiro Oda
天野喜孝	Yoshitaka Amano
大友克洋	Katsuhiro Otomo
新海诚	Makoto Shinkai
葛饰北斋	Katsushika Hokusai
竹久梦二	Takehisa Yumeji
皆叶英夫	Hideo Minaba
空山基	Hajime Sorayama
盐田千春	Chiharu Shiota
村上隆	Takashi Murakami
细田守	Mamoru Hosoda
草间弥生	Yayoi Kusama
大友良英	Otomo Yoshihide
铃木敏夫	Toshio Suzuki
荒川弘	Hiromu Arakawa
岸本齐史	Masashi Kishimoto
/ 中国 /	
顾恺之	Gu Kaizhi
吴道子	Wu Daozi
阎立本	Yan Liben
王维	Wang Wei
李思训	Li Sixun
荆浩	Jing Hao
董源	Dong Yuan
巨然	Ju Ran
黄荃	Huang Quan
徐熙	Xu Xi
范宽	Fan Kuan
李公麟	Li Gonglin
米芾	Mi Fu
钱松岩	Qian Songyan
何香凝	He Xiangning
陈之佛	Chen Zhifo
李苦禅	Li Kuchan
张书旂	Zhang Shuqi
王雪涛	Wang Xuetao
蒋兆和	Jiang Zhaohe
叶浅予	Ye Qianyu
赵少昂	Zhao Shaoang
董希文	Dong Xiwen

艺术家	
赵佶	Zhao Ji
张择端	Zhang Zeduan
李唐	Li Tang
刘松年	Liu Songnian
马远	Ma Yuan
夏圭	Xia Gui
赵孟頫	Zhao Mengfu
黄公望	Huang Gongwang
王蒙	Wang Meng
倪瓒	Ni Zan
吴镇	Wu Zhen
沈周	Shen Zhou
唐寅	Tang Yin
仇英	Qiu Ying
董其昌	Dong Qichang
徐渭	Xu Wei
陈洪绶	Chen Hongshou
八大山人	Bada Shanren
石涛	Shi Tao
郑板桥	Zheng Banqiao
任伯年	Ren Bonian
吴昌硕	Wu Changshuo
齐白石	Qi Baishi
黄宾虹	Huang Binhong
徐悲鸿	Xu Beihong
张善子	Zhang Shanzi
高剑父	Gao Jianfu
高奇峰	Gao Qifeng
张大千	Zhang Daqian
刘海粟	Liu Haisu
颜文梁	Yan Wenliang
潘天寿	Pan Tianshou
林凤眠	Lin Fengmian
黄君璧	Huang Junbi
黄少强	Huang Shaoqiang
吴湖帆	Wu Hufan
胡佩衡	Hu Peiheng
林散之	Lin Sanzhi
常玉	Chang Yu
傅抱石	Fu Baoshi
陈少梅	Chen Shaomei
吴作人	Wu Zuoren
李可染	Li Keran
刘奎龄	Liu Kuiling